This is (not) Rocket Science

How a New Generation of Core Independent Peripherals

Redefined Embedded Control

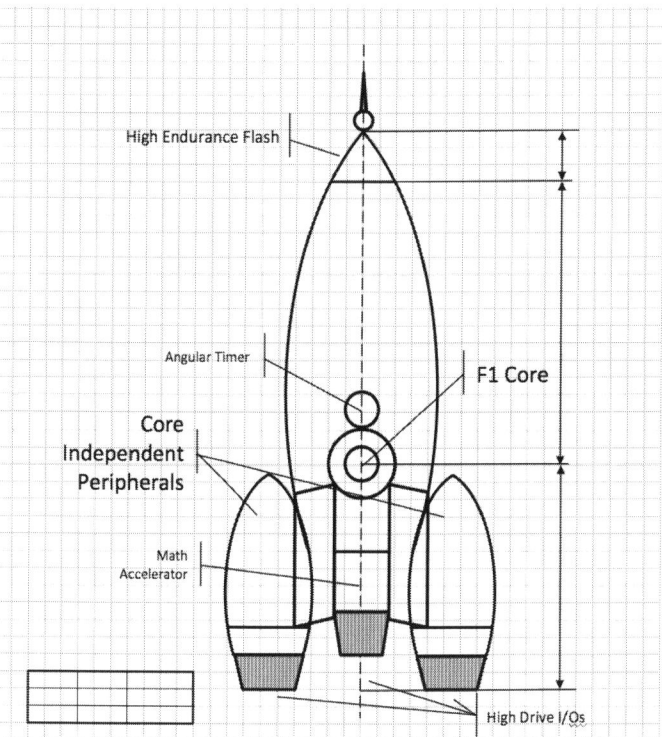

Lucio Di Jasio

All rights reserved
No part of this book may be reproduced, scanned, or distributed in any printed or electronic form.

Copyright © 2015 by Lucio Di Jasio

Visit us at: http://www.flyingpic24.com

E-mail: pilot@flyingpic24.com

An acknowledgment and thanks to Lulu Enterprises, Inc. for making the publishing of this book possible.
http://www.lulu.com

ISBN: 978-1-312-90777-5

Printed in the United States of America

First Edition: 2015

This book is dedicated to

Dieter

Table of Contents

Acknowledgments 1

Introduction 3
   It is Not About The Core 3
   Core Independent Peripherals 4
   Rocket Science 6
   How To 7
   Who Should Read This Book 7
   Software Tools 7
   Hardware Tools 8
   What This Book is Not 8
   Online Support 9

Chapter 1 - Preflight Checklist 11
   Installing MPLAB X 11
   MPLAB XC8 Compiler 12
   Installing MPLAB XC8 13
   Installing MPLAB Code Configurator 14
   Creating a Project 15
   Hardware Prototyping 26

Chapter 2 - Timing Functions 31
   Oscillators 31
   8 and 16-bit Timers 35
   CCP – Capture Compare and PWM 39
   CWG – Complementary Waveform Generator
   COG – Complementary Output Generator 44
   NCO – Numerically Controlled Oscillator 48
   HLT – Hardware Limit Timer 53
   SMT – Signal Measurement Timer 57

Chapter 3 - Input / Output 61
   I/O Ports 61
   CLC – Configurable Logic Cells 64
   PPS – Peripheral Pin Select 68
   DSM – Data Signal Modulator 71
   ZCD – Zero Cross Detect 73

Chapter 4 - Non Volatile Memory 77
   Data EEPROM 77
   Flash Memory 78
   HEF – High Endurance Flash 79

Chapter 5 - Safety Functions 83
   CRC – Cyclic Redundancy Check with Memory Scanner 83
   WDT – WatchDog Timers 86
   Reset Circuits 88

Chapter 6 - Communication Functions 91
I2C – Inter Integrated Circuit Bus 91
SPI – Synchronous Port 94
EUSART – Asynchronous Serial Port 98
USB – Universal Serial Bus – Active Clock Tuning 102

Chapter 7 - Analog Functions 107
Comparators 107
DAC – Digital to Analog Converter 110
FVR – Fixed Voltage Reference 113
ADC – Analog to Digital Converter 116
Temperature Indicator 122
OPA – Operational Amplifiers 124

Chapter 8 - Math Support 127
AT – Angular Timer 127
PID – Math Accelerator 131

Chapter 9 - eXtreme Low Power 135
XLP – eXtreme Low Power 135
Low Power Modes 136
PMD – Peripheral Module Disable 138

Acknowledgments

Against my best efforts even this book turned out to take way more pages and time than initially planned. Once more, it would have never been possible for me to complete it if I did not have 110% support and understanding from my wife Sara. Special thanks go to Greg Robinson, Sean Steedman and Cobus van Eeden for reviewing the technical content of this book and providing many helpful suggestions.

I would like to extend my gratitude to all my friends, colleagues at Microchip Technology and the many embedded control engineers I have been honored to work with over the years. You have so profoundly influenced my work and shaped my experience in the fantastic world of embedded control.

Finally thanks to all my readers, especially those who wrote to report ideas, typos, bugs, or simply asked for a suggestion. It is partially your "fault" if I embarked in this new project, please keep your emails coming!

Introduction

Ten years have passed since I started writing my first embedded C programming books, focusing first on 16-bit and eventually 32-bit architectures. At the time, like most others in this industry, I was predicting a slow but sure decline of the 8-bit applications. Instead, 8-bit PIC® microcontrollers have continued to grow steadily with volumes that have well passed the mark of the billion (that is 1 followed by nine zeroes) units shipped per year, and keep growing.

Much to my own surprise, and perhaps right because of the competitive pressure presented by the new generations of low end 32-bit microcontrollers, it is in the 8-bit PIC architecture that I have witnessed the most spectacular and exciting innovation happen. This started with the evolution of the *PIC16F1* core followed by the gradual introduction of the *core independent peripherals*.

It is Not About The Core

The adoption of the C programming language has accelerated in the past decade to become the norm today in all embedded control applications. All the while, assembly programming has been relegated to a few specialized applications and, even there, often isolated to small sections within an application.
The development of the so called **Enhanced Mid Range core**, or more simply what I call the PIC16 **F1 core** from the easily recognizable first digit used in all parts featuring it, was the natural response from Microchip 8-bit architecture team. The *original* PIC® core had been optimally designed for the most efficient use in hand coded assembly, but was notoriously challenging for compiler designers. The architects of the 8-bit PIC quickly squeezed in a few more instructions (twelve in total) to help with the *code density* of typical C compiler constructs and, most cleverly, provided it with a new *linear access* mode to memory, removing the number one offender, RAM banking, from the compiler designer's concerns. While at it, they also dropped in an automatic interrupt context-saving feature and for good measure doubled the hardware stack so to make interrupt handling and in general complex function nesting more agile.

The new core enhancements were introduced so smoothly that most of the "legacy" users hardly noticed the core transplant! In fact the implementation allowed for total backward compatibility with decades of mostly assembly applications. while, in new applications, the C compiler abstracts away all the underlying architectural detail and the user experience relates only to the improved speed, interrupt response capabilities, increased stack depth, and code compactness.

Core Independent Peripherals

But the true focus of all recent innovations has been the *peripheral set* of the 8-bit PIC microcontrollers with the introduction of the **Core Independent Peripherals** or **CIPs** for short. This introduction represents a major shift in the way PIC microcontroller solutions are designed today and most importantly in how they differ from the approach of the low end 32-bit MCUs that are competing for the same applications space. Where the latter are assuming an ever stronger focus on software (more of it, meaning more complexity and requiring higher clock speeds and more power), the Core Independent Peripheral approach is the exact opposite. It is by focusing on autonomous and directly interconnected hardware peripheral blocks that 8-bit applications can achieve *more* while reducing software complexity, delivering faster response times at lower clock speeds and using less power!

Three Steps to (Embedded Control) Zen

As in oriental disciplines, reaching the *illumination*, getting it, is a gradual process of interior growth. Truly understanding the Core Independent *philosophy* requires at least three major steps:
1. *Discovery*, when the disciple is introduced for the first time to a new way of approaching the challenges of this (embedded) world. Large sections of this book are devoted to presenting brand new core independent peripherals and/or enhancements to familiar ones. For each new CIP I will attempt to explain the target use cases and hint at possible further applications to achieve specific core workload reductions.
2. *Mastery*, where the disciple starts understanding the deeper philosophy behind the design of the new CIPs and comprehends how they can be interconnected, like LEGO® blocks, to create new *functions* and to

change radically the overall application design. Each chapter in the book is dedicated to a specific function that can be better expressed by combining one or more of the peripherals presented.
3. *Zen*, where the disciple takes a holistic approach, freely mixing analog and newly constructed digital functions to remove most/all hard timing constraints from an application, reducing code size and complexity by achieving *automatic* responses via *chains* of (hardware) events to create *complete* (mixed signal) *solutions*.

Why the CIPs Paradigm is Different

Individually taken, CIPs are not necessarily absolute novelties. Most all of them are or have been available in other, often competing, architectures or as part of larger systems. What is different though is their focus on *autonomous operation* and *direct interconnection*.

Further some readers will be tempted to compare the assembly of CIPs to create new *functions* with programmable logic (FPGA) system design. But the difference in the approach is vast when you consider the granularity of the building blocks. The CIP philosophy consists in re-assembling really large functional units (complete peripherals) rather than individual gates. Despite the apparent loss in flexibility the CIP approach has three major advantages:

- The *skill set* required is of the kind that is most available in the embedded control industry. Any engineer with even minimal exposure to firmware development, but also most computer science graduates, electrical engineers and even hobbyists, will be immediately familiar with the CIPs. This is definitely NOT the case when programmable logic is used for a complete SoC design to include (soft) core development, where the skill set required is highly specialistic including the familiarity with hardware description languages, adding specific core architecture and interfacing deep knowledge, possibly augmented by complex proprietary model libraries.
- The *power consumption* of a pure programmable logic application is orders of magnitude higher. Microcontrollers featuring the CIPs are capable of eXtreme Low Power operation.
- The *cost* of a pure programmable logic application is also orders of magnitude higher. Microcontrollers featuring CIPs are offered at the lowest end of the price range.

Where Can I Find the Core Independent Peripherals

In typical Microchip conservative approach, new core independent peripherals have been added gradually to every new family of PIC16F1 microcontrollers introduced over the last five years or so. This was done for two major reasons: to reduce risk and to keep the new microcontrollers competitive. In fact, just like in software development, when introducing a new peripheral, possibly replacing an older module, there is always the danger of introducing new bugs. A too fast proliferation of a module, before it is fully debugged and understood, is a sure way to cause a lot of trouble. At the same time, adding too many features to a product without a clear (application) focus means only increasing the cost without a clear benefit to the end user.

In the appendix A you will find a snapshot of the most current Peripheral Integration Guide (DS30010068), a document that you can find on Microchip web site (updated at least quarterly), that summarizes the current CIP allocation/availability across the many new microcontroller families.

Rocket Science

The introduction of the Core Independent Peripherals poses a new set of challenges. The new paradigm requires the embedded control designer to make a small step outside his comfort zone. As the saying goes: "if you have a hammer in your hand every problem looks like a nail". So if the triplet (timer, ADC, PWM) is where the comfort is, the natural approach to any new application is to throw more of the above at the problem and patch it all up with sufficient amounts of software, interrupts and above all MIPS.

Add to that the dominant, perhaps Moore low induced, culture that wants *more* (software/performance) to equate with *better* and you can appreciate the magnitude of the effort required to change perspective.

So the challenge is on you, my reader, to decide to take a look at what new tools are available, and dare going against the crowds, looking for core independent solutions that require *less* software, *less* complexity and *less* power. This is NOT Rocket Science!

Introduction - 7

How To

This (short) book is modeled after a typical (8 hours) lecture and I expect you to spend approximately the same amount of time to read it all through, although possibly not all at once. It takes time to appreciate some of the subtle changes introduced and it is important that you plan on practicing/testing manually the material presented. I strongly encourage you to follow my hardware and software setup (first chapter) and to play and experiment with at least some of the CIPs, perhaps starting with those that appear to relate most with your particular field of expertise and personal interests. Each module, offers a "homework" section where I will invite you to reflect on the possible implications and possibilities offered by the new capabilities. A large number of references to online material will also be offered to help you continue deepening your understanding of the matter.

Who Should Read This Book

- If you think you know 8-bit microcontrollers, think again, these are not your grandfather's microcontrollers anymore!
- If you knew and liked PIC microcontrollers, but have not looked at them in a while, you will get a kick out of this!
- If you think you knew PIC microcontrollers but hated their cryptic instruction set and many idiosyncrasies, you should take a second look now!
- If you have been trying to use small 32-bit microcontrollers, but got put off by the rapidly escalating complexity and the excessive reliance on software when trying to realize even the simplest real time functions, you will get a refreshing new perspective!

Software Tools

We will use exclusively **MPLAB® X**, a free and cross platform (Windows, OS X and Linux) integrated development environment (IDE), and the free **MPLAB® XC8 compiler**. We will also make extensive use of the **MPLAB® Code Configurator (MCC),** a plug-in that will allow us to generated a safe and space-optimized API (a set of functions written in C) to initialize and control the microcontroller peripherals for the *specific* configuration required by our application.

Please note that **all such tools are living projects** and as such they could and will eventually diverge from the version I used while writing this book (early 2015).

Hardware Tools

In order to provide a hands on learning experience, all the demo projects in this book can be tested using (free) samples of a few models of PIC16F1 microcontrollers in readily available DIP packages mounted on any standard bread board. A Microchip Universal In-Circuit Debugger/Programmer **PICkit™ 3 (PG164130)**, despite the extremely low cost, will allow you to program and debug more than 1,000 different PIC microcontroller models regardless of their (8, 16 or 32-bit) core architecture. You can also consider a **PICDEM™ Lab Development Kit (DM163045)**, bundling breadboard and PICkit 3 programmer/debugger, or perhaps most cost effectively, the recently introduced **PICDEM™ Curiosity** board featuring an integrated programmer/debugger and novel modular expansion options (Click boards).

What This Book is Not

This book is not an introduction to Embedded Programming, or a primer in C programming. This book assumes already a *basic* level of C programming expertise and some previous knowledge about microcontrollers technology.

This book does not replace the individual PIC16F1 microcontroller datasheets, in fact I will often refer the reader to such material for further study. Similarly this book cannot represent a comprehensive summary of all the features offered by the PIC16F1 microcontrollers used and/or the tools used.

Should you notice a conflict between my narration and the official documentation, ALWAYS refer to the latter. However, when you do so, please remember to send me an email (at pilot@flyingpic24.com), I will publish and share any correction and/or useful hint on the blog and book web site.

Online Support

All the source code developed in this book is made available to all readers on the book web site at: *http://www.flyingpic24.com*. This includes additional (bonus) projects and a complete set of links to *online code repositories* and third party tools as required and/or recommended in the book.

Over the last few years I have been contributing to a blog, "The pilot logbook", at *http://blog.flyingpic24.com* and I will continue to do so time permitting.

Chapter 1 - Preflight Checklist

In this chapter we will prepare our initial software and hardware setup. This will include, downloading and installing:

- ***MPLAB® X*** (v2.26 or later), if you are still rooting and/or using MPLAB 8, just let it go! There are simply too many good reasons to upgrade to list them all here.
- ***MPLAB® XC8*** (v1.33 or later), this C compiler is common to all 8-bit devices.
- **MPLAB® Code Configurator** plug-in (v. 2.10 or later), this tool will help us quickly generate optimized C functions to access all old and new (core independent) peripherals available on PIC16F1 models.

By the end of this chapter we will develop the embedded world equivalent of the classic "Hello World" program, blinking an LED connected to one of the device digital I/O pins. Further examples (short snippets) will be offered in the following chapters right after each new peripheral introduction to illustrate the ease of configuration achieved by the combination of the above three development tools.

Installing MPLAB X

If you have already installed and previously used the MPLAB X IDE and the XC8 compiler, you can skip this section and jump directly to the Code Configurator plug in installation.

Otherwise follow the link below to go straight to the MPLAB X support and download page on Microchip web site:

> http://www.microchip.com/mplabx

Here is what's important to remember when you install MPLAB X:
1. Make sure to select the right version for your OS. Check your browser download window and verify that you are getting an *.exe*

file if you are using Windows, a *.dmg* file for OS X or a *tar.gz* file for a Linux box.

2. If you had a previous version of MPLAB X installed, remember that you have to **uninstall it** first! You can trust MPLAB X un-installer, your projects, tools and configurations will be remembered and will be restored safely as soon as the new version is up and running.

3. On the contrary, there is no need to uninstall previous versions of MPLAB 8. MPLAB X will NOT overwrite or corrupt previous MPLAB 8 installations.

Online Resources

http://microchip.wikidot.com – Microchip Technical Training Team has developed a relatively large *wiki* that includes many short step-by-step tutorials and a good number of videos too. It's well worth browsing through it, check out the FAQs and some nifty tips and tricks.

MPLAB XC8 Compiler

Contrary to MPLAB X IDE, a tool that covers all PIC® microcontrollers, MPLAB XC8 is specific to 8-bit PIC microcontrollers.
Note that MPLAB XC8 is offered in two versions:

- Free version, which offers only a basic set of optimization options

- *PRO version*, which includes the Omniscient optimization technology providing superior code density and/or speed

Note that the Free version includes also a time limited (60 days) PRO license. So you can have a taste of what those additional optimization modes can buy you and decide if the saving is worth the price based on the benefit to your specific application and coding style.

For the purpose of this book, as in my previous, we will NOT need any of the advanced features of the optimizer and the Free version will suffice to give us more than adequate performance. In fact, most applications will run perfectly satisfactorily without using even the most basic of the optimization options.

Installing MPLAB XC8

From the MPLAB X download web page, you will also find links to download the correct version of the MPLAB XC8 compiler for your operating system.
After download, launch the installer and you will be presented with a short sequence of dialog boxes:

1. A classic click-through license agreement. You **must accept** the legal terms and conditions if you want to continue.

2. You will be offered a choice between installing the XC8 compiler, a Network License server or updating your license settings. Choose to **install the XC8 compiler** at this time.

3. Next, you will be asked what kind of XC8 compiler license you plan on using. Unless you are part of a large organization where a Network Licensing scheme is in use, choose to **install the XC8 compiler on your computer** (local access key).

4. At this point you will be confronted for the first time with the *License Activation Manager,* which will prompt you to enter your license activation key. Unless you happen to have one such key, simply **leave the field empty** and **click on Next**.

5. Eventually **click on Yes** in the final confirmation dialog box (see Figure 1.1)

6. Lastly, you will be asked if you want to activate the time limited PRO *evaluation period*. Note that this feature can be turned on at a later time, when most convenient to you. So don't worry about it for now, **run the compiler in Free mode**. Learn to use the tool first and develop your code. When ready, you will be able to launch the *License Manager* (XCLM tool) and start your 60 days evaluation period.

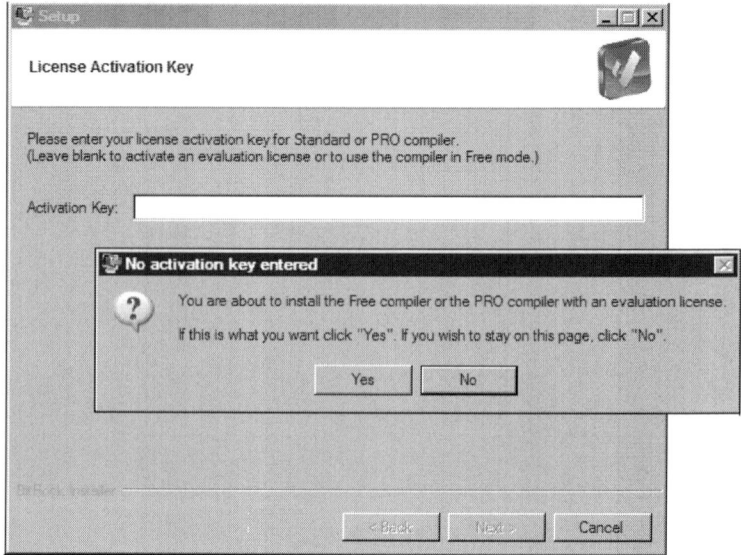

Figure 1.1: License Activation Manager dialog

Installing MPLAB Code Configurator

MPLAB Code Configurator is a plugin of MPLAB X, and as such it can be installed and activated from within the application.

From the main menu select **Tools > Plugins** and then from the **Available Plugins** pane select **MPLAB Code Configurator** (see Figure 1.2) and click **Install**.

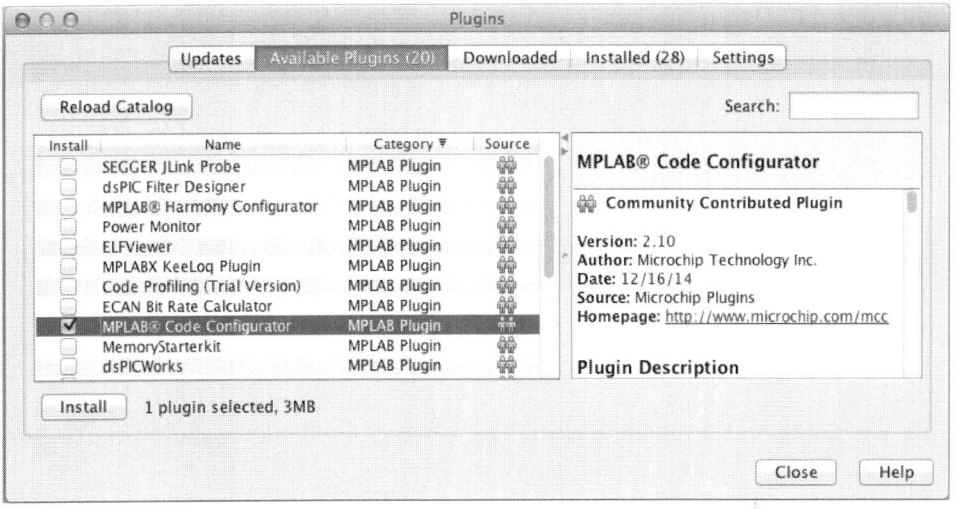

Figure 1.2: Plugins dialog box

Once installed, you will be asked to restart MPLAB X before the first use of the tool so that the MCC can be added to the Tools > Embedded menu. (Figure 1.3)

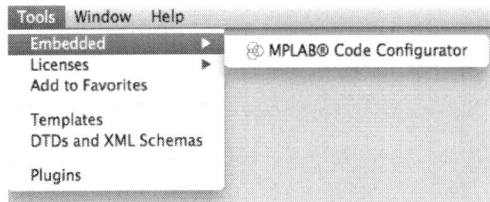

Figure 1.3: MPLAB Code Configurator Activated

Creating a Project

If you have already used MPLAB X and the XC8 compiler to create a project, you can skip this part as well and jump directly to the Prototyping Setup section below.

New Project Checklist

This simple step-by-step procedure is entirely driven by the MPLAB X *New Project* wizard.

From the **Start Page** of MPLAB X, select **Create New Project**, or simply select **File>New Project...** from the main menu, it will guide us automatically through the following seven steps:

1. *Project type selection*: in the *Categories* panel, select the **Microchip Embedded** option. In the *Projects* panel, select **Standalone Project** and click **Next**.

2. *Device selection*: in the *Family* drop box, select **Mid-range 8-bit MCU**. In the *Device* drop box, select the desired PIC part number. In this particular example we will select the **PIC16F1708,** and click **Next**.

3. *Header selection*: this step is skipped automatically, you don't need one with this PIC family, but other part numbers (especially for very small pin count packages) might offer you the option.

4. *Tool selection*: select **Simulator**, and click **Next**.

5. *Plugin board selection*: this step is skipped automatically.

6. *Compiler selection*: select **XC8** (v1.33 or above), and click **Next**.

7. *Project Name and Folder selection*: type "**1-HelloWorld**" (no spaces) as the project name, then browse to your working directory and click **Finish** to complete the wizard setup.

After a brief moment, you will be presented with a new *Projects* window (see Figure 1.4). This will be empty except for a small number of *logical folders*.

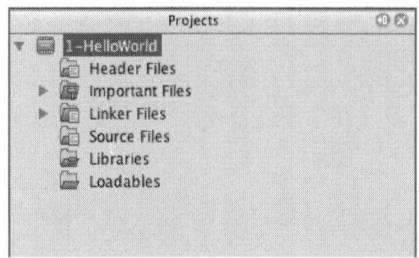

Figure 1.4: Project Logical Folders

Preflight Checklist - 17

So what is a "project" in MPLAB X? If you fire up the File Explorer (Finder for MAC users) and navigate to the working directory, you will find that MPLAB X has created a folder with the name you assigned to the project and appended the extension *.X* to it. Think of this folder as THE project!

NOTE FOR MPLAB 8 EXPERTS

> Since an MPLAB X project is effectively a folder, double clicking on it won't automatically launch MPLAB X, but simply will tell your file manager to inspect the directory contents. This is a sort of disappointment for old time MPLAB 8 users. If you are aching for some drag and drop action though, you can drag a project directory from your file manager into an MPLAB X (open) window.

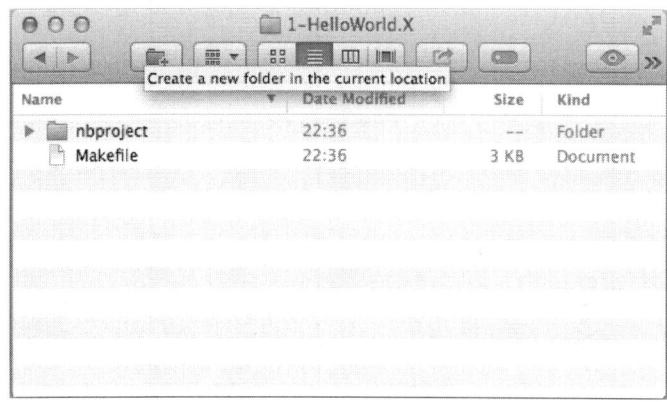

Figure 1.5: MPLAB X Project Folders

You will notice how the project folder (see Figure 1.5) is not completely empty; there are always at least two elements in each MPLAB X project:

- *Makefile*, this is exactly what it claims to be, an (automatically generated) make file that will be used by the GNU *make* tool to build your project
- *nbproject* folder, this is where MPLAB X stores the configuration of your project, including the list of source file names to be compiled, your personal preferences, debugging tools selections and so on.

NOTE

> The name of this folder is revealing the origin of MPLAB X. It has never been a mystery that MPLAB X is based on the NetBeans IDE project.

Needless to say, you don't want to mess with the contents of these two folders as both are generated and maintained automatically by MPLAB X.

If you stick to some basic guidelines, that I will highlight in the following, you can make sure that the entire folder contents are *position independent*. This means that you will be able to move (or copy) the entire project folder to a different path on your hard drive or on a different machine and the project will remain intact and fully functional.

Logical Folders

So what is the relationship between the *logical folders* (Figure 1.4) and the contents of the actual physical folder (Figure 1.5)?

It is actually much more loose than you think. The *logical folders* are simply lists of file names. The location of those files is independent of the actual position of the project folder we just created.

The most important logical folder is the one named: *Source Files*. ALL and ONLY files listed in here will be compiled and linked into our applications regardless of their location in the file system.

On the contrary, files listed in the *Header Files* logical folder, are there merely for our convenience. You can leave this folder/list empty if you want and your project will compile just fine! Obviously, it pays to be disciplined and list in here the most important header files in a project. I recommend you do so to help better document and maintain the project.

Similarly the *Important Files* folder contains just a link to the makefile. It is there for your reference. Feel free to inspect its contents with the built in editor (double-click on the file name), but don't try to modify it by hand. Adding other files to this folder won't make a difference to the project build process.

The *Linker Files*, *Library Files* and *Loadable Files* folders will not be used in this book and I will defer the explanation of their purpose to the MPLAB X official documentation.

New File Checklist

Time to create the *main* project source file. There are at least three ways to add a new file and in particular a *main.c* file to a new project:

1. Invoke the *New File Wizard*, and use a template.
2. Invoke the New File Wizard, and start from scratch.
3. Launch MPLAB Code Configurator and let it generate it automatically (recommended, if you are lazy as I am).

The *New File Wizard* can be activated by the **CTRL-N** command (⌘-N for MAC users), selecting **File>New File** from the main MPLAB X menu or by clicking on the *New File icon* on the **File Toolbar** (if active).

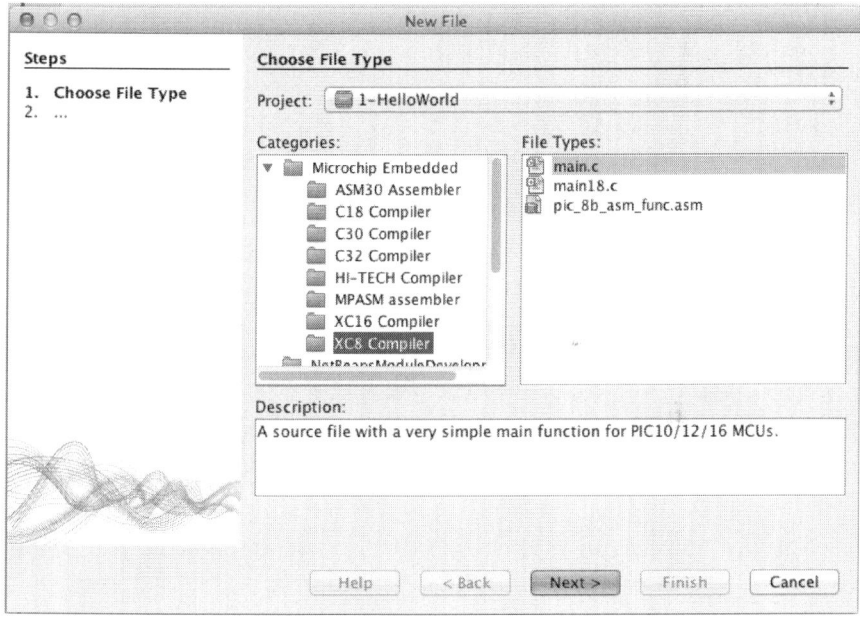

Figure 1.6: New File Wizard

The New File Wizard (see Figure 1.6) is composed essentially of two dialog boxes and requires the following steps:

1. *File Type selection*: In the *Categories* pane, select **Microchip Embedded**.
2. This will expand into a list of sub-categories, select **XC8 Compiler**.
3. In the right pane titled *File Types*, select the **main.c** type.

4. Click **Next**.

5. The *Name and Location* Dialog box will appear. Here most fields will be already pre-filled with the default settings of your project (folder), you will have only to assign a proper name to the new file, type: **main.c**

6. Click **Finish**.

MPLAB X will create the new *main.c* file from the xc8 specific template that is composed of the following few lines of code:

```
/*
 * File:    main.c
 * Author:  (your name here)
 *
 * Created on (date and time here)
 */

#include "xc.h"

void main( void )
{
    return;
}
```
Listing 1.1 – MPLAB XC8 *main.c* Template

The alternative, perhaps more common case of use of this Wizard when not creating a *main.c* file, is to create an empty file and then type your way through it. In this case, in the first dialog box you will choose the **Other** category and the **Empty File** type.

Note that, in both cases, the wizard not only creates the file and populates it with a template as required but, if you had the *Source Files* logical folder open and selected in the Project window, it does also automatically **add the newly created file to the project**.

If that did not happen, after saving the new file, you will have to manually select the **Source Files** folder in the project window and activate the context menu to select **Add Existing Item...**

As mentioned above, the third option to create a new *main.c* file for our projects is to let MPLAB Code Configurator do it for us!

Launching the MCC

The MPLAB Code Configurator is most helpful if used from the very beginning of a project setup. You can activate MCC by selecting **Tools>Embedded** from the MPLAB X main menu as seen in Figure 1.3.

Should you find the Embedded menu empty or missing, you will have to verify that the MCC plugin has been installed correctly and/or that MPLAB X has been closed and restarted after the installation to activate the new plugin.

When successfully activated MCC will present itself (see Figure 1.7) as the combination of two (eventually three) dialog windows: the *resources selection* window to the left, the *configuration dialog* window overlapping as a tab the editor windows, and later the *Pin Manager* window that will take a right column position.

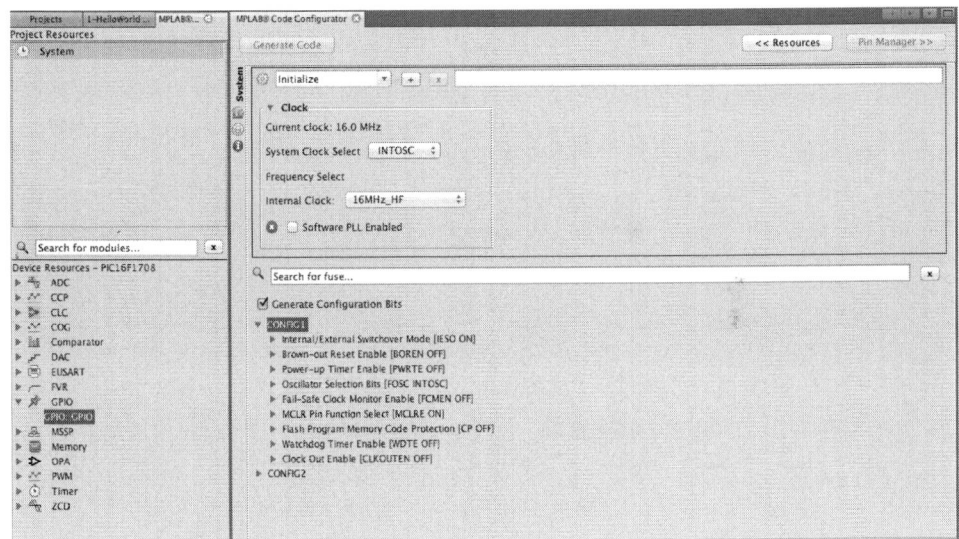

Figure 1.7: MPLAB Code Configurator, System Dialog Window

The resources selection window is further split vertically in two panes. The top one, will list the resources that we select to include and use in our project. The bottom pane lists all the resources available (and yet not used) on the chosen microcontroller model.

System Settings

Initially the *Project Resources* list (top left pane) contains only a single resource named: *System*. **Click on it** to activate the corresponding configuration dialog box in the central dialog window. From there we can set the default system configuration at power up as follows:

- System Clock select: **INTOSC**

- Frequency Select: **4 MHz** (leave the PLL unchecked for now)

Checking the **Generate Configuration Bits** option (bottom half of the screen) set the following options that will be convenient during the development phase:

- **Config1** word: all options off, with the exception of **MCLRE ON**.

- **Config2** word: all options off, with the exception of ZCDDIS OFF and PPS1WAY OFF.

The following chapters will reveal the nature of the specific features and the reasons of such choices.

At this point the main MCC dialog window will show that the Generate Code button is active (showing a 1 in brackets) to indicate that at least one module configuration (System) has changed and can now be (re) generated. **Click** on it and the MCC will start the code generation process.

In doing so for the first time, MCC will also check if a *main.c* file is already present and (if you chose the third, lazy, option before) if missing, will offer to create one for you (see Figure 1.8).

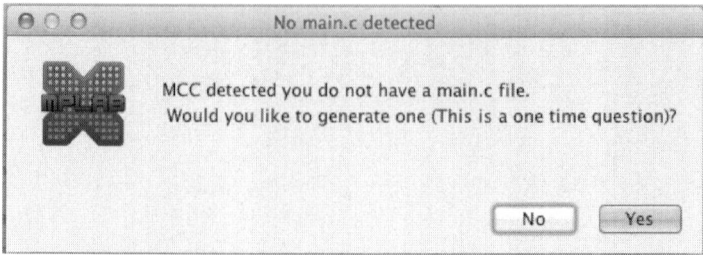

Figure 1.8: *main.c* **Automatic Generation**

After this first run, the MCC will add a couple of new elements to our project:
- A *main.c* file if this was missing (and you approved)
- An MCC Generated Files logical folder, inside which..
- The *mcc.c* file containing the SYSTEM_ and OSCILLATOR_Initialize() functions in addition to the Configuration bit settings (#pragmas).
- In parallel, the corresponding header file (*mcc.h*) will be created and added to the Header Files logical folder

Managing I/Os

In order to get our "Hello World" message out to the world, we will also need to activate at least one I/O pin. We can ask MCC to help us setting all the required configuration details by selecting the GPIO (General Purpose I/O) resource from the Device Resources list (bottom left panel).

Double-click on GPIO and when it appears in the Project Resources list (top left panel) **click once more** on it.

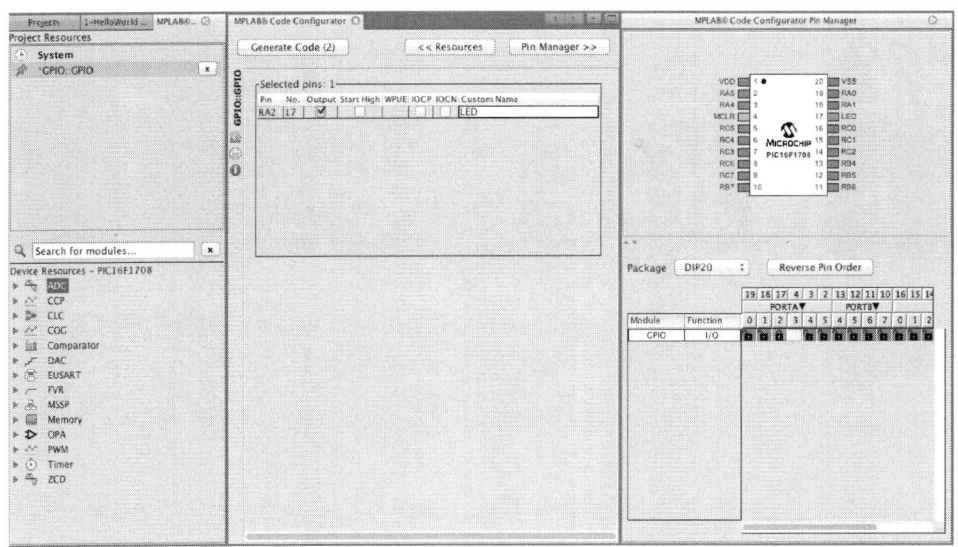

Figure 1.9: MCC, GPIO Dialog Window and Pin Manager Window

This changes the configuration dialog window to show the Selected pins table and activates the Pin Manager – new window appearing to the right as seen in Figure 1.9.

In the Pin Manager window (bottom right table), **select** the little blue lock corresponding to the RA2 pin, turning it green and closing the lock.

In the main dialog window you will see RA2 being added to the table of selected pins.

Now **check the Output** option, and **click on the Custom Name** field (rightmost) to edit the name, let's call it: **"LED"**.

Finally **click** once more on the **Generate Code** button to let MCC add the new GPIO module and settings to the project.

A new file called *pin_manager.c* (and corresponding header file) has been added to the list of MCC Generated Files. Inside it you will find the default `PIN_MANAGER_Initialize()` function that will take completely care of all the required I/O control registers initializations.

Even more interestingly, the new *pin_manager.h* file contains already a number of macros (`#define`) that manage our custom named pin such as: `LED_SetHigh()`, `LED_SetLow()` and `LED_Toggle()`.

Hello Simulated World

Every respectable programming book must contain a "Hello World" example. In the embedded world this is not necessarily done using text on a screen/terminal, but often simply giving an indication of activity by means of an LED blinking.

Up to this point we have not written (manually) a single line of code, yet our project contains already a well structured and complete set of functions (an API) ready to initialize the microcontroller with the desired power up options, oscillator and reset configuration and ready to manipulate the LED output.

So let's get to it by editing the main loop and adding just a single line of code to the template `main()` function that MCC generated for us:

```
/*
   Generated Main File
   ...
*/
#include "mcc_generated_files/mcc.h"

/*
   Main application
 */
void main(void)
{
    SYSTEM_Initialize();

    while (1)
    {
        LED_Toggle();
    }
}
```
Listing 1.2 – Hello (Simulated) World

To test the code we can now **build the project** and **run** it using the *MPLAB X Simulator*. From the main menu you can select **Run>Run Project**, or use the keyboard **F6** shortcut (MAC users will have to press **Fn+F6**).

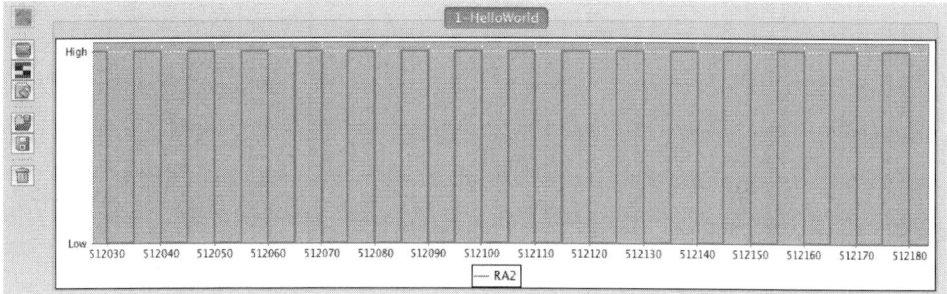

Figure 1.10: Simulator Analyzer window

After **opening the *Simulator Analyzer*** window, we will **pick the RA2 pin** output and visualize it as in Figure 1.10.

It might take a little zooming to be able to distinguish the actual square wave output produced on pin RA2, but that is to be expected as we have been toggling the pin as fast as the processor could run through the main loop.

Hardware Prototyping

The MPLAB Simulator can be useful for a quick check of basic application logic. Most of the basic peripherals are supported by the simulator and inputs can be injected as *synchronous or asynchronous stimuli*. But many of the new core independent peripherals are actually much more entertaining to test if we can use a little hardware prototyping.

Breadboard + PICkit™ 3

The most "basic" option consists in simply plugging a DIP packaged microcontroller on a breadboard and providing the five basic In-Circuit Serial Programming™ (ICSP™) connections to a PICkit 3 programmer and debugger (see Figure 1.11 for an example of the proper wiring).

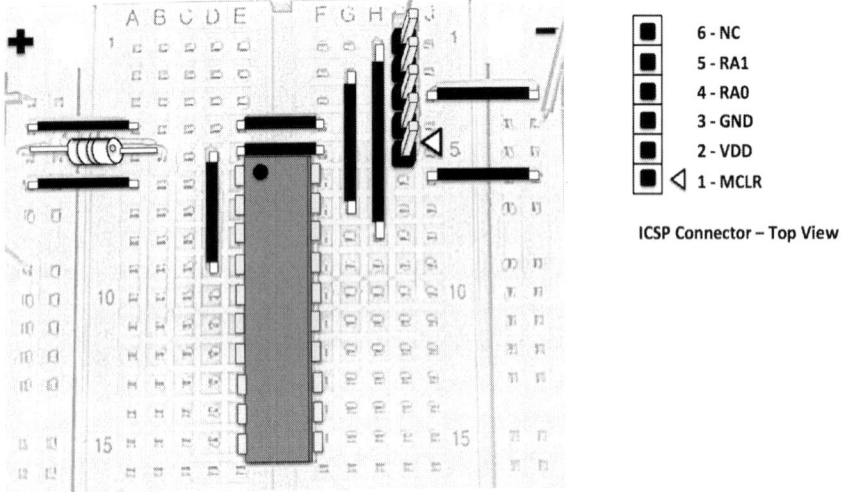

Figure 1.11: Hardware Prototyping using a Breadboard

Note that the PICkit 3 can also provide the power supply to the target circuit if configured so. Figure 1.12 illustrates the *Project Configuration* dialog box where the PICkit 3 has been selected and the *Power settings pane* has been chosen to enable the *Power Target* function and 5V output.

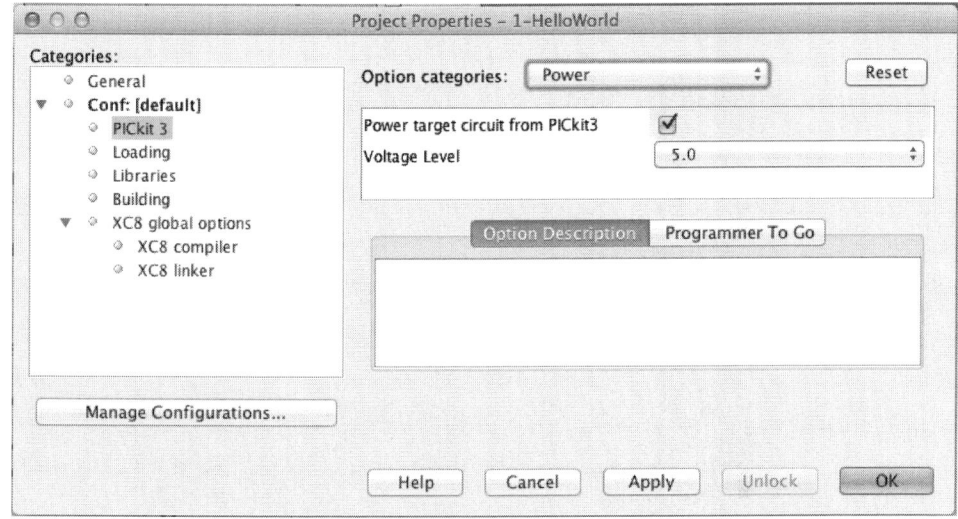

Figure 1.12: Configuring the PICkit3 to provide power

PICDEM™ Lab Development Kit

The PICDEM Lab Development Kit (DM163045) bundles a PICKit3, a breadboard and a number of sockets where the connections to the PIC microcontroller are already wired for you.

Low Pin Count Demo Board

An alternative option for quick prototyping with the low pin count (up to 20-pin) PIC16F1 models consists in using one of the *Low Pin Count Demo Boards* (DM164130-9) that can also be bundled with the PICkit3 as part of the *PICkit3 Starter Kit* (DV164130). Note that in such a case the LEDs populated on the board are connected to the RC0-RC3 pins. You will need to change the GPIO configuration accordingly but our first code example will otherwise work unmodified.

PICDEM™ Curiosity

Figure1.13: PICDEM Curiosity board

The latest and greatest prototyping option is represented by the **PICDEM Curiosity board** (see Figure 1.13). This board provides various power supply sources and connects directly (via USB) to MPLAB X for programming and debugging without need for an external device (no PICkit3 required). It features also (right top corner) a convenient *mikroBUS*™ connector to access over a hundred *click*™ *boards,* an original product series of MikroElektronika D.O.O. (*http://mikroe.com*) providing sensors, actuators, displays and a seemingly infinite selection of wired and wireless interfaces.

Hello Real World

Regardless of the hardware prototyping solution of your choice, simply **connect** a small resistor (500 – 1k Ohm) in series with an LED directly between the chosen output pin of the PIC16F1708 and ground.
After adding a (plain and dirty) half second delay to the main loop (as shown in listing 1.3), **rebuild the project** and **run** it on the target. From the main menu you can select **Run>Run Project**, or use the keyboard **F6** shortcut (MAC users will have to press **Fn+F6**)

```
/*
   Generated Main File
   Device: PIC16F1708
*/
#include "mcc_generated_files/mcc.h"

/*
   Main application
 */
void main(void)
{
    SYSTEM_Initialize();

    while (1)
    {
        int i;
        for( i=0; i<32000; i++);    // delay approx 1/2 sec
        LED_Toggle();
    }
}
```
Listing 1.3 – Hello (Real) World

If all went well, you should now be able to see the LED blinking at a rate of approximately 1 Hz.

Hello Real World!

Homework

Looking back at the steps we followed to generate this first project you should realize the amazing amount of work that MCC has done for us.

- Inspect visually all the files that have been generated.
- Notice how MCC has documented our selections when loading configuration registers.
- Modify the configuration (System clock or GPIO selections for example) and let MCC regenerate the source files. Observe the changes.
- Manually modify some of the source files generated and then have MCC regenerate them. What happens when there is a conflict?
- Add a second GPIO initialization function. (Hint: use the plus button next to the Initialize combo box).

Chapter 2 - Timing Functions

Oscillators

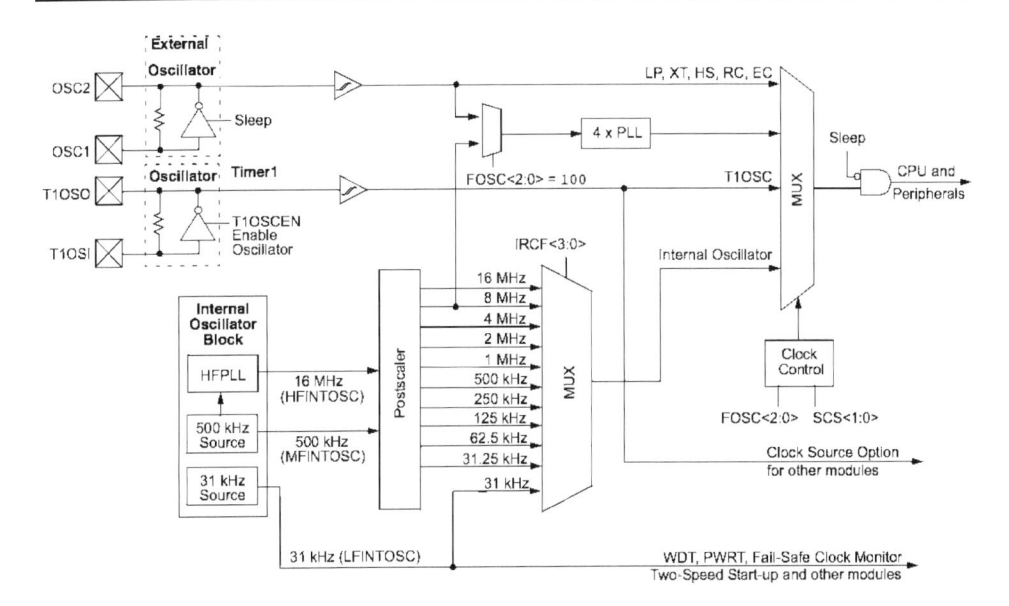

Figure 2.1: Simplified Clock Source Block Diagram

Introduction

It might come as a surprise to some, but most/all recent PIC microcontrollers are featuring at least five (5) different on-chip oscillator circuits. These include:

- *Primary External Oscillator* (POSC): Its typical use is as a system clock source in applications that require tight timing tolerances (typically 1% or better) for asynchronous serial communication. The circuit gain can be adjusted to optimize power consumption and crystal performance depending on frequency by selecting one of three possible modes: (LP, XT, HS). Further, EC mode provides a direct input when a suitable clock signal is already available.

- *Secondary Oscillator* (SOSC): It can be connected to a second crystal directly tied to a 16-bit timer (Timer1) independent from the main system clock. It is a low gain circuit dedicated to 32KHz operation to perform the role of a low-power Real Time Clock (RTC).

- *Internal Oscillator* (HFIntOSC): It is typically used in alternative to the Primary External Oscillator in low cost applications or in those cases were particularly tight tolerances are not needed. This circuit can operate at different frequencies to optimize power consumption and provide a selection of frequencies from 31kHz to 32MHz.

- *Low Power Internal Oscillator* (LFIntOSC): This is a low-power and relatively low-accuracy oscillator that is used by the Watchdog circuit but also by several accessory modules including BOR, Fail Safe Clock and PowerOn Timer circuit.

- *ADC Oscillator* (FRC): This is a low-power internal oscillator of fixed frequency (600 kHz nominal) that is dedicated to the exclusive use of the Analog to Digital converter module and allows it to operate while the rest of the microcontroller is in Sleep mode (see XLP in chapter 9)

Additional dedicated oscillators circuits and/or modes are offered by selected peripherals modules (see the USB or the PSMC module for examples).

How They Work

The internal oscillator (HFIntOSC) actual tolerance depends on the particular model and will vary with ambient temperature and supply voltage although at nominal values ($T_{ambient}$ and 5V) it is factory calibrated to deliver typical 1% accuracy. The presence of a Tuning register (OSCTUN) makes it possible to implement further (custom) calibration techniques at the end of the production line or even run-time *compensation* for temperature and voltage provided either or both can be accurately measured (see voltage reference FVR and Temperature Indicator).

Since external crystals require a settling time before they can provide a accurate and stable clock signal, the *Oscillator Start-up Timer* (OST) is automatically enabled and forces a delay of 1024 counts when one such mode is selected. To minimize the latency between external oscillator start-up and code execution, a *Two-Speed Start-up* mode can be selected. This is a device configuration option that allows the device to power-up quickly and begin immediately executing application code using an internal oscillator and switch to the external (accurate) oscillator later, once stable and ready.

For those cases when the power supply (Vdd) is rising very slowly, it is instead possible to extend the power-up sequence by enabling an automatic (64ms) long delay via the *Power-up Timer* (PWRT) to ensure Vdd reaches a suitable level before code execution begins. A PLL circuit (x4) is available on models that operate up to 32MHz in order to extend the upper frequency range of the internal and external oscillators. The PLL can be controlled in software or can be configured to run from power-up. Two-Speed Start-up, Power-up Timer and PLL Enable are device Configuration Bits options.

Applications

Oscillator tolerances and power consumption characteristics define the range of their applications:

- Asynchronous Serial Communication (UART) requires <2% tolerance, mandating the use of external crystals/resonators when the temperature and/or voltage range cannot be limited/compensated for.
- USB communication (Full Speed 12MBs) requires <0.25% tolerance, requiring the use of external crystals or *Active Clock Tuning* (see Chapter 6)
- Battery applications typically require the lowest operating frequencies and lowest power modes of the internal/external oscillators. Also the use of internal oscillators provides the fastest start up time with potential large savings in the overall application power consumption.

Limitations

A few entry level models of recent introduction (such as the low pin count models of the PIC16F150x family) do not offer the full complement of external oscillator modes to help control cost and complexity.

MCC API

The MCC groups most of the oscillator configuration options in the *System* resource. Through the same dialog window it is possible to generate the device *Configuration Bits* that are responsible for the oscillators configuration at power up (FOSC), enable a *Clock Output (CLKOUTEN)* feature, enable the *External/Internal* clock switch capability (IESO), Fail Safe Clock Monitor (FCMEN) and configure the Watchdog (see an example in Chapter 1 – System Settings).

The MCC places the resulting Configuration-Bits (`#pragma`) and the `Oscillator_Initialize()` function in the *mcc.c* file. The initialization function call is automatically inserted in the `SYSTEM_Initialize()` sequence.

PinOut

When enabled the following features will make exclusive use of some of the I/O pins of the microcontroller overriding any other setting:

- CLKOUT (out) — clock output ($F_{osc}/4$), for cascading applications or testing/debugging purposes
- OSC1 (in) and OSC2 (out) — connect to external crystal/resonators
- OSC1 (in) — connect to external clock sources (EC modes)
- T1OSI (in) and T1OSO (out) — connect the low-power secondary oscillator 32KHz crystal (typically available on 28-pin devices and larger)

Homework

- Learn about clock-switching techniques to run your applications and peripherals at the optimal clock speed at any point in time.
- Learn about Two Speed start up and see how it can help your application be "productive" quicker and save time/energy.
- Check the Fail Safe Clock monitor feature (more on this in Chapter 5 - Safety functions).

Online Resources

AN1303 - Software Real-Time Clock and Calendar
AN1798 - Crystal Selection for Low-Power Secondary Oscillator
AN849 - Basic PICmicro® Oscillator Design

8 and 16-bit Timers

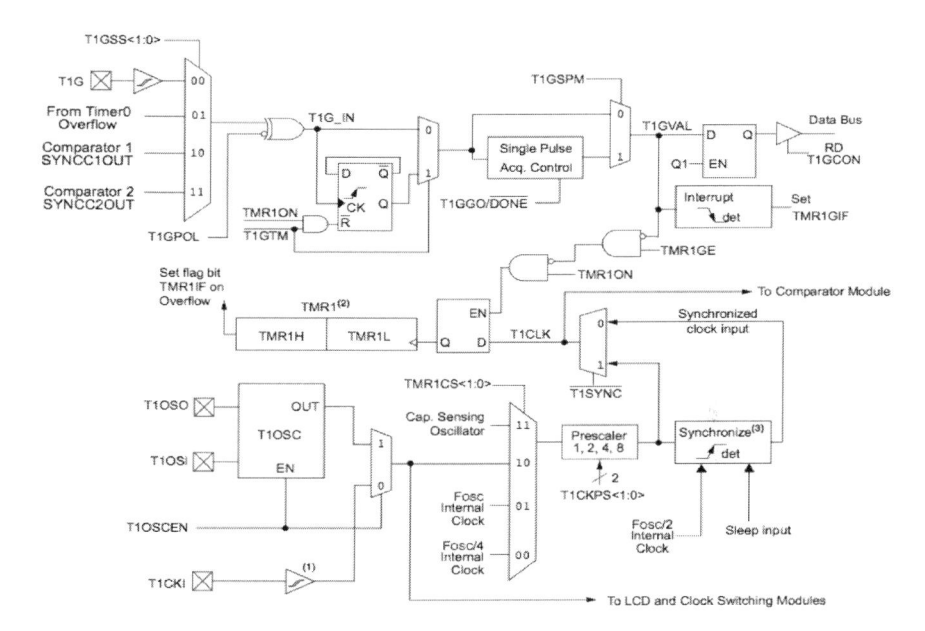

Figure 2.2: 16-bit (Odd) Timer Block Diagram

Introduction

The fact that Microchip designers are so obsessively focused on providing backward compatibility among all PIC microcontrollers is never more obvious than when analyzing the structure of the timers section.

There are essentially three types of timers:

- Odd numbered timers, such as Timer1, Timer3, Timer5 ..., are slightly more advanced 16-bit structures that can operate in combination with the secondary oscillator SOSC module (see Chapter 2. Oscillators section) as well integrate with the Capture and Compare (see Chapter 2 - CCP section).

- Even numbered timers, such as Timer2, Timer4, Timer6..., are simple 8-bit structures featuring a pre-scaler, a post-scaler and a simple period register. Notice their clock source is pretty much limited to the system (instruction) clock Fosc/4.

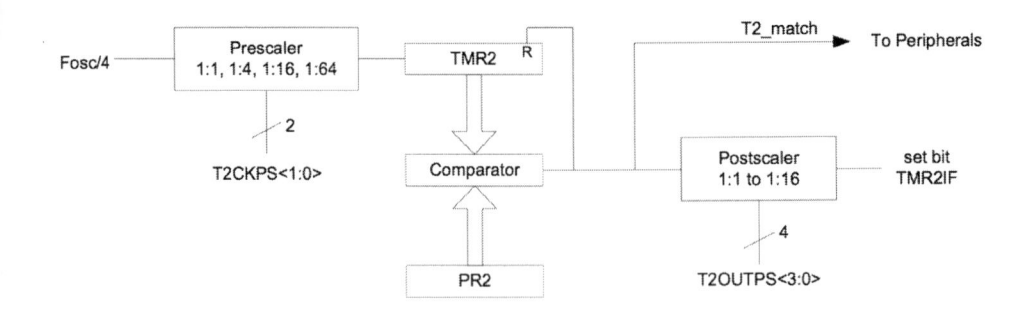

Figure 2.3: 8-bit (Even) Timer Block Diagram

- Timer0, is in a class of its own. It is an 8-bit structure of the simplest possible design and provides the ultimate backward compatibility link with the previous 25 years of PIC application history. Notice the absence of a period register but the ability to operate as a counter for pulses presented on the T0CKI input pin.

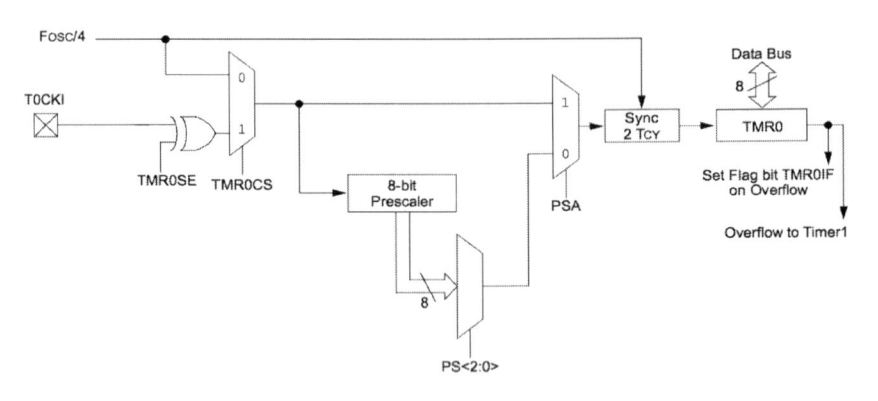

Figure 2.4: Timer0 Block Diagram

How They Work

All timers regardless of their size and type provide at least a basic period generation capability (timer mode) and the ability to operate as a pulse counter (counter mode) via the input TxCKI pin.

Nowadays, Timer0 use is relatively limited as, missing a comparator and a period register, it can only generate multiples (256x) of the clock period unless CPU cycles are spent to detect the counter rollover (TMR0IF) and to reload manually an *offset*.

Even (8-bit) timers are more flexible and require only setting the corresponding period (PRx) registers with the desired values. In fact when used with the CCP and PWM modules they are responsible for controlling the PWM signal period.
Odd (16-bit) timers are the most flexible of the group and can operate as gated Timers, but also, will cooperate with the CCP module to perform pulse width measurements (capture mode), or to produce single pulses of desired length (compare mode).

Applications

Needless to say that the applications of these timers are the most disparate, including: scheduling events, producing safety timeouts, measuring input signal frequency, period, and/or duty cycles...
It must be noted that the 8-bit timers (Timer2 in most cases) are also occasionally offered as a baud rate generator option in some communication peripherals (see I^{2}C / SPI module).

Limitations

Despite the relatively advanced capabilities of the 16-bit timers, certain common functions such as measuring an incoming PWM signal duty cycle or producing a fixed on-time PWM output can be deceivingly complex and demand heavy use of CPU cycles. For such applications we will see how the recently introduced Signal Measurement Timer (SMT) and/or the Hardware Limit Timer (HLT) will come to our rescue.

MCC API

The MCC offers a dedicated dialog window for each timer module present on a given device and will produce a corresponding timerX.c file containing optimized TimerX_Initialize(), Start/Stop and Read/Write functions for each.
If a timer *interrupt* option is enabled, MCC will also automatically create a corresponding entry in the *Interrupt Manager* vector table (and will emit the *interrrupt_manager.c* file).
As a special case, enabling interrupt handling for Timer0 will allow MCC to add a *period reload* feature simulating (at a cost of some CPU overhead) the capabilities of a more complete Even (8-bit) timer.

PinOut

The early PIC16F18xx and PIC16F19xx families did provide only fixed positions for each of the timers gating and clock inputs (T1G, T1CKI...).
The low cost PIC16F15xx family added the ability to connect directly timer inputs via the Configurable Logic Cell to a number of alternate pins and to other peripheral outputs.
More recent devices (PIC16F170x, PIC16F171x, PIC16F188xx ...) allow the Peripheral Pin Select module to further connect the timer inputs (and outputs) to any available digital I/O.

Homework

- Investigate asynchronous operation of the 8 and 16-bit timers.
- Investigate Timer1 Gate mode.

Online Resources

TB3100 - Timer1 Timer Mode Interrupt Latency

CCP – Capture Compare and PWM

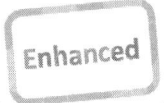

Enhanced

Introduction

The Capture Compare and PWM module (CCP) has been part of the PIC microcontroller peripheral set for the most part of the last 25 years. As such its design shows the *frugality* that was typical of those early days when every single gate was chosen and placed by hand and nothing could be wasted.

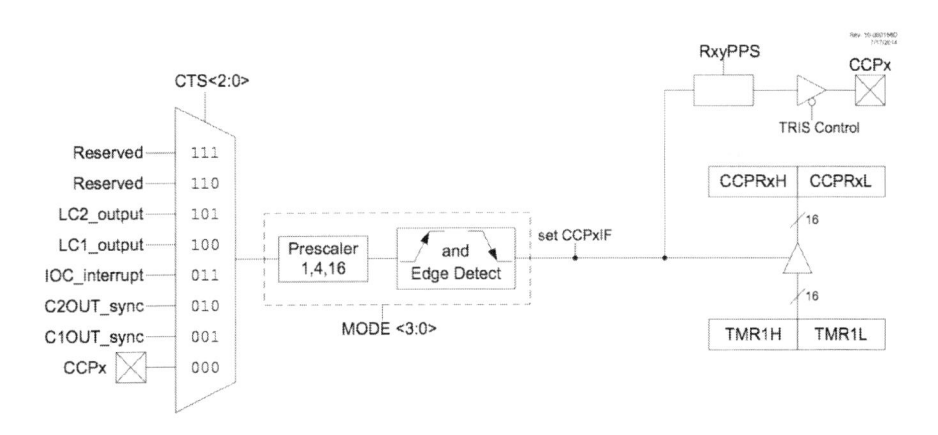

Figure 2.5: Capture Mode Operation Block Diagram

In fact to perform two of its most basic operations: capture and compare, it relies on the connection to one of the 16-bit timers (see Figure 2.5). Consequently the resolution of the capture and compare functions is 16-bit, which provides a relatively large dynamic range.

To produce a PWM output the CCP module relies instead on the connection to an 8-bit timer and its period register (see Figure 2.6).

The resulting period resolution is therefore limited to only 8-bit, and the duty cycle would be similarly constrained if it was not for a clever use of the full system clock. While all other functions of the microcontroller are based on the instruction clock of the processor (Fosc/4), the CCP (in PWM mode) has access to the full system clock (Fosc) adding 2 more bits to the count for a grand total of ten.

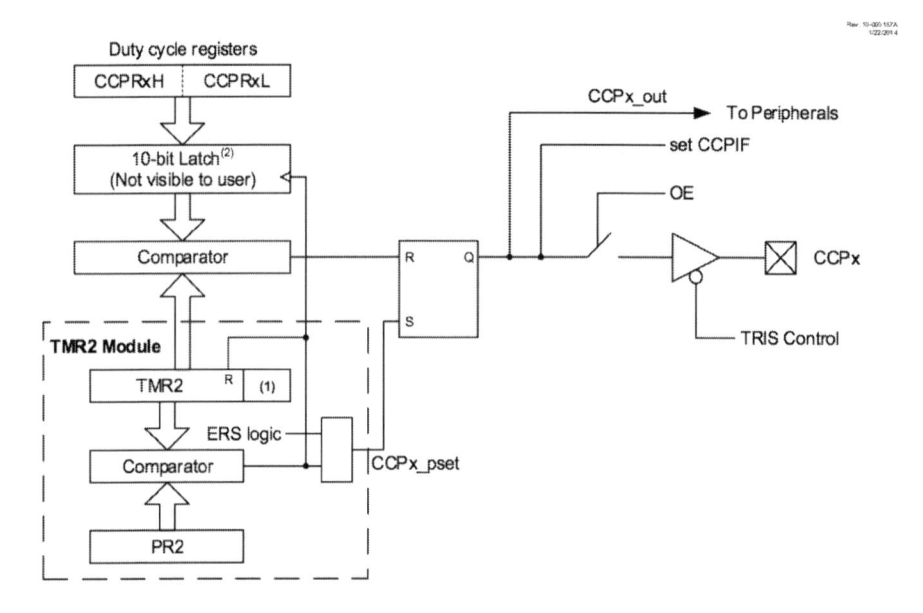

Figure 2.6: Simplified PWM Block Diagram

How it works

As a consequence of the design limitations presented above, each CCP module needs either an 8-bit or a 16-bit timer to perform one of its functions.
A CCP module plus an odd (16-bit) timer will be able to perform a 16-bit *capture* measuring the width of an incoming pulse.
A CCP module plus an odd (16-bit) timer can also be used to perform a 16-bit *compare* function, producing an output pulse of desired width.
A CCP module plus an even (8-bit) timer will be able to produce an output *PWM* signal with 8-bit period resolution and 10-bit duty cycle resolution.
On selected modes (PIC16F19xx for example) the PWM mode is further *enhanced (ECCP)* to produce a pair of complementary output signals instead of a simple pulse/PWM square wave. This is convenient in many power applications where a pair of complementary MOSFET devices is eventually being connected to a load. Dead-band control (delay) can be added to prevent the simultaneous activation of both power devices during a commutation (and the resulting shoot-through current) due to the large gates parasitic capacitance.
In more recent designs though the portion of the circuit responsible for the *complementary output control* has been de-coupled from the CCP module and

made instead available as a completely separate module. (see Chapter 2. CWG, COG).

PWM Only Modules

Similarly, in recent microcontroller families, additional modules capable of the sole PWM functionality have been added to the mix of peripherals in acknowledgment of the fact that this was by far the most common use case of the CCP module.

Applications

Pulse measurement/generation (capture/compare) are extremely common in the most varied range of embedded applications to interface to sensors and control simple actuators.

PWM outputs are used to control servos, they are common in power supply and most motor control applications.

They can also be seen as a digital to analog conversion tool. A simple RC filter is in fact sufficient to turn any PWM output in an analog output signal with up to 10-bit of resolution.

Limitations

Despite the nominal resolution (number of bits available in the Duty Cycle comparator circuit), the *effective (DC)* resolution of a PWM is always limited by the clock source available:

a)
$$F_{osc} = 2^{PWMresolution} * PWM_{freq}$$

b) solving for the PWM resolution:

$$PWM_{resolution} = \log_2(F_{osc} / PWM_{freq})$$

Formula 2.1 - Effective PWM resolution

As illustrated in Formula 2.1b, for a given clock frequency (F_{osc}) as the PWM frequency is increased the effective resolution of the resulting PWM decreases. Assuming a 32MHz system clock, for example, it can be demonstrated that the maximum resolution of 10-bit is attainable only for PWM frequencies < 32KHz.

MCC API

The MCC provides a great simplification in the use of the CCP and PWM modules as it automates completely the process of selecting and connecting the right timers to the CCP modules. If enabled, interrupt vectors will also be added to the *interrupt manager* module.

Beside the default `CCPx_Initialization()` function, the MCC will produce different APIs depending on the selected function:

- Capture: a convenient `CCPx_IsCapturedDataReady()` allows polling for the capture event and `CCPx_CaptureRead()` returns the pulse/event duration.

- Compare: a `CCPx_CompareCountSet()` allows to set the desired pulse output width, while the `CCPx_IsCompareComplete()` function allows to poll for the pulse generation sequence completion.

- PWM: the `PWMx_LoadDutyValue()` function simplifies setting the 10-bit Duty Cycle register.

PinOut

Early PIC16F18xx and PIC16F19xx models have traditionally provided fixed input/output pin selections for the CCP/PWM modules.

More recently, PIC16F150x families have added the capability to re-route all CCP-PWM pins to the CLC module and through it to a larger selection of I/Os and peripherals.

Most recent families (PIC16F16xx and PIC16F17xx for example) have added the Peripheral Pin Select (PPS) feature and with it the ability to route the CCP=PWM signals to any available digital I/O.

Homework

- Investigate how to control a servo using a CCP or PWM module

- Investigate how a PWM can be used to produce an analog output and contrast it with the use of a DAC

- When can multiple PWM outputs share the same timer and when do they need separate ones?

Online Resources

AN1175 - Sensorless Brushless DC Motor Control with PIC16

AN1261 - Dimming Power LEDs Using a SEPIC Converter and MCP1631 PWM Controller

AN1305 - Sensorless 3-Phase Brushless Motor Control with the PIC16

AN1562 - High Resolution RGB LED Color Mixing

AN594 - Using the CCP Module

Example

The following example illustrates how a CCP or PWM module can be used to control a Servo motor:

```
/* Project: ADC to PWM Servo
 * Device:  PIC16F1509
 */

#include "mcc_generated_files/mcc.h"

#define TCLK    _XTAL_FREQ / 4
#define TPERIOD  (unsigned char)(TCLK/4 * 0.008)  // 125Hz period (8ms)
#define SERVO_MIDDLE   (unsigned)(TCLK * 0.0014) // 1.4ms
#define SERVO_MIN      (unsigned)(TCLK * 0.0004) // 0.4ms
#define SERVO_MAX      (unsigned)(TCLK * 0.0024) // 2.4ms

void main(void)
{
    // configure ADC to trigger from Timer2 and generate an interrupt
    // configure PWM1 for an 8ms period
    SYSTEM_Initialize();

    // Enable Interrupts
    INTERRUPT_GlobalInterruptEnable();
    INTERRUPT_PeripheralInterruptEnable();

    while (1)
    {
    }
}
```

```c
/* edited in the adc.c file
*/
void ADC_ISR( void)
{ // read potentiometer value and translate to servo angle
    uint16_t duty;

    duty = SERVO_MIN + ( ADC_GetConversion( Potentiometer));
    if ( duty > SERVO_MAX)
        duty = SERVO_MAX;
    PWM1_LoadDutyValue( duty);

    // Clear the ADC interrupt flag
    PIR1bits.ADIF = 0;
}
```

CWG – Complementary Waveform Generator
COG – Complementary Output Generator

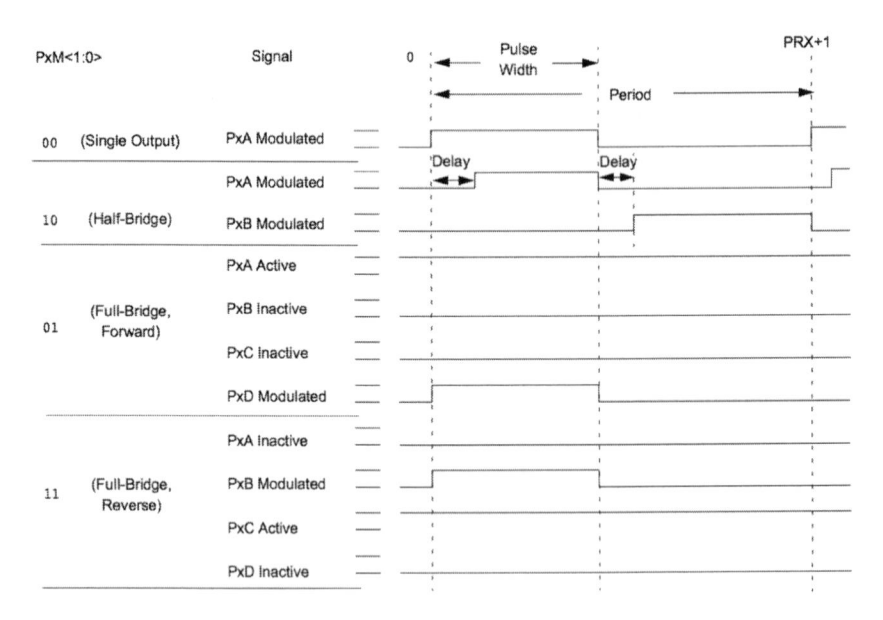

Core Independent

Figure 2.7: Complementary Output

Introduction

While it is true that many motor control applications require a PWM output signal to be split in two complementary waveforms (and adding a dead band delay as in Figure 2.7) in order to drive a pair of MOSFET power transistors,

Timing Functions - 45

this is not an exclusive relationship. There are many more applications in power supply and lighting that can benefit from the same "service". This realization has driven the PIC16F1 architects to separate what was the back-end of the ECCP module into an independent module on its own called Complementary Waveform Generator (CWG) and later the more advanced Complementary Output Generator (COG).

How it works

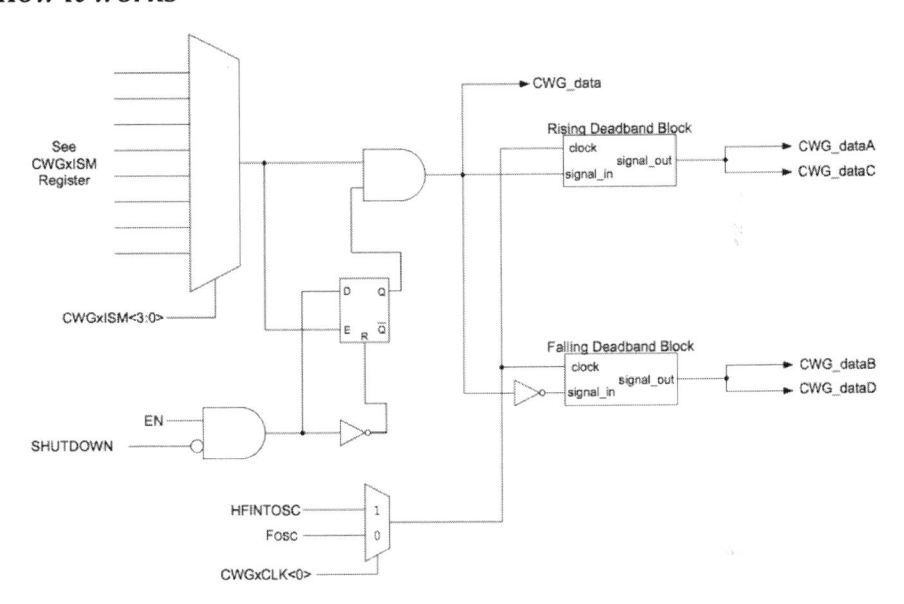

Figure 2.8: Simplified CWG Block Diagram (half bridge)

The CWG module as illustrated in Figure 2.8 performs all the functionalities of a typical ECCP back-end and some, but most importantly, thanks to the large input selector multiplexer (or ISM in the architects jargon) it can now be connected with any number of other signal generating modules among which: NCOs, high speed comparators, CLC blocks, and of course 10 and 16-bit PWMs. This makes it a much more useful and versatile tool in the designers hands. In other words: "the total is now greater than the sum of the parts".
A CWG module can perform the following functions:
- Independent (dead band) delays on the rising and/or falling edges accounting for the different size (and gate capacitance) of N and P MOSFET devices, therefore allowing further optimization of the circuit efficiency.

- Automatic shutdown control
- Output steering control
- Half bridge and Full bridge output modes
- Push Pull output modes
- Can operate even during Sleep, provided independent event sources are selected (Note: if HFIntOSC is used as the main clock of a COG, its oscillator will NOT be disabled during CPU sleep!)

A COG module can perform the following additional functions:
- Add an input blanking counter to cancel multiple commutations should the input events be "noisy".
- Add a phase delay to help stabilize the control system.
- Accept (multiple) separate inputs for rising and falling event sources.
- Accept level or edge sensing input.
- Asynchronous dead-band delay chains are offered in alternative the usual synchronous counter in order to provide the finest granularity (down to 5ns).

Applications

While a comprehensive list of applications would be impossible, here are some of the most distinctive applications supported by CWG and COG modules:
- Push-Pull topologies are found in selected power supply applications
- Half Bridge and Full Bridge modes are used to control brushed DC motors with optional direction control.
- High Speed comparators connected as sources of rising or falling edge events are commonly used in power supply (switching) applications (histeretic, peak current mode...).
- NCO output connected with a Half Bridge driver can be used to design High Intensity Discharge lamps and Dimmable Fluorescent lamp ballasts.

Limitations

Because of the integrated automatic shutdown, blanking and complex output modes provided, the CWG/COG module simplifies considerably the design of power supply and motor control applications. It removes the need for CPU support during each cycle and therefore the need for fast interrupt responses or other (CPU intensive) workloads. The main limitations in all above applications is often represented by the available maximum PWM clock source which is usually constrained by the device maximum clock speed (commonly 32MHz). Still, it is common to see power supply applications achieve switching frequencies in excess of 400kHz while using only a small fraction of the CPU available performance.

For applications that require a higher resolution, see the PSMC module, at the time of this writing available only on the PIC16F178x family, which is capable of PWM operation from a dedicated 64MHz oscillator.

MCC API

The MCC groups conveniently all the CWG/COG options in four panes containing: Output Pin Configurations, Events control, Auto-Shutdown and Steering control.

Beside the usual CWGx_Initialize() it generates a minimal API containing three groups of functions:

- Functions to optimize the individual dead-band counters.
- Functions to set or clear the Auto-Shutdown event.
- A function to load (all) new settings at once, synchronizing the switch between modes helps avoiding dangerous (as in costly) conflicts when driving high power MOSFET devices.

PinOut

The presence of the Peripheral Pin Select feature on all the new device families gives them the most absolute flexibility on the choice of the optimal pinout. The only exception to this is when the designer wants to take advantage of dedicated high drive strength pads (see Chapter 2- 100mA Output drives for more details).

Homework

- Investigate possible uses of the CWG/COG deadband control to introduce automatically short time delays in a chain of events.
- Investigate how the CWG/COG could be used to "double" the frequency of an input signal.

Online Resources

TB3118 - Complementary Waveform Generator Technical Brief
TB3119 - Complementary Output Generator Technical Brief
TB3120 - Slope Compensator on PIC Microcontrollers
AN1660 - A Complete Low-Cost Design and Analysis for Single and Multi-Phase AC Induction Motors Using an 8-Bit PIC16 MCU
AN1779 - Sensored Single-Phase BLDC Motor Driver Using PIC16F1613

NCO – Numerically Controlled Oscillator

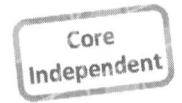

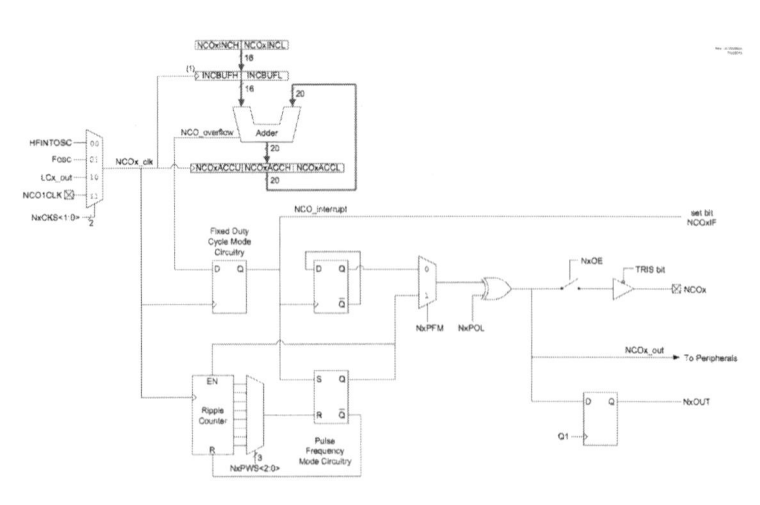

Core Independent

Figure 2.9: NCO Block Diagram

Introduction

The Numerically Controlled Oscillator represents another twist on the timer/counter concept. Where normal timers increment their count by one at each clock pulse, the NCO can instead increment by an arbitrary number of steps. In Figure 2.9 is represented by the easily recognizable V shape of an adder block. Further, where other timers reach their period match or maximum count, they reset and start counting from zero. An NCO instead simply produces a carry bit, that is used to generate an output pulse, and continues counting up using the reminder. The consequences of such differences might not be obvious but has very important implications that make this module very interesting for a number of applications in power supplies and more specifically in lighting applications.

How it works

Since the timer does not start every period counting up from the same value (zero), but keeps building on the reminder of the previous period, it follows that some periods are going to be (one clock tick) shorter than others. It follows that the frequency output of the timer is not going to be constant, but rather oscillating between two near values cycle by cycle. The distribution of shorter and longer cycles will be such to average out to a very precise value that we can demonstrate to converge rapidly to an exact ratio of the input clock given by the formula:

$$F_{out} = F_{in} * Increment / 2^{20}$$

Formula 2. - NCO Averaged Output Frequency

Since the Increment register value shows up at the numerator of the fraction, we deduce that there is a *linear relationship* between it and the resulting output frequency value (Fout). This is a very important property that makes the NCO stand apart from all other timers/counters we have seen so far.

Studying further the NCO mechanics we discover that especially when the output frequency is very high (I.e for large values of the Increment) each further increase of the register produces a proportionally very small increment in the output frequency (1/Increment is a very small number if Increment is big). In other words the *frequency resolution of the NCO is maximum when the output frequency is the highest*. Once more this is the opposite behavior compared to a tradition timer where at maximum output frequency corresponds the most coarse frequency control.

As a final observation, since the NCO produces only a carry bit as its output, eventually used to toggle an output pin or to produce a single fixed length pulse, the *output duty cycle is essentially fixed and independent of the frequency*. Once more, this is the opposite of the traditional timers/PWM modules where maintaining a 50% output or a constant DC with the changing frequency requires additional CPU support.

In conclusion, it might be worth reviewing a few special cases for the Increment value, they are:

- Increment is set to 1, the NCO becomes now a normal counter, in fact a very large one (20-bit).

- The Increment register is set to a value that is a power of two (i.e. 2^N), where the result is equivalent to a traditional timer whose input clock has passed through a prescaler/divider and the resulting frequency is now Fout = Fin * 2 $^{(N-20)}$.

In all such special cases the NCO basically reverts to the behavior of a traditional timer, and as such we might consider using it when its peculiar characteristics are not required by the application at hand.

As is always the case for Core Independent Peripherals, the function performed by the NCO can be otherwise performed with a more traditional timer module and software by sufficient application of additional CPU workload. When used, the NCO can eliminate completely such CPU overhead. Additionally the NCO has proved to be quite versatile and has been successfully used in combination with CLC blocks and traditional timers to construct new and more complex custom peripherals.

Applications

Summarizing the differences between a traditional timer/PWM module and the NCO in Table 2.1. we can observe how the two modules are almost the exact opposite of each other.

	NCO	PWM
Frequency	Linear control	Fixed
Duty Cycle	Fixed	Linear control
Frequency Output	Linear	Fixed
Resolution at Fmax	Maximum	Minimum
Output Jitter	1 (tick)	None

Table 2.1 - Comparison of NCO and Timer/PWM

The NCO can be advantageous in many applications where a natural inertia of the system exist, be it in the form of mechanical inertia, capacitance of an output circuit or resonance of the circuit. In all such cases the NCO higher resolution and linear response can transform a complex control problem by providing a much simpler and inexpensive solution.

In particular the properties of the NCO have been proven of great advantage to many power supply applications such as High Intensity discharge lamp ballasts and Dimmable Fluorescent lamps to control very accurately the circuit current (after ignition) by moving with high (frequency) resolution down the side of the resonant circuit response curve.

Limitations

Note that while the original NCO module had a 20-bit adder and a 16-bit increment, which implies that the maximum output frequency is limited to 1:16 of the input clock.

More recent models (as found in the PIC16F171x family) have increased the Increment register size to 20-bit so that a 1:1 maximum ratio is achievable expanding considerably the output frequency range.

MCC API

The MCC support for the NCO is mostly concerned with helping compute the proper Increment values and to estimate the output frequencies range achievable for the given clock input.

The minimal API in the generated *nco1.c* source file provides only the default NCOx_Initialize() function and, if interrupts are not enabled, a status polling macro: NOCx_GetOutputStatus().

PinOut

Early PIC16F150x models offered only a fixed position (RA5) for the NCO external input (NCO1CLK) and up to two alternate options for the NCO output (RC1 and RC6) in 20 pin devices.
More recent PIC microcontroller families featuring the Peripheral Pin Select function allow any available digital I/O pin to be used.

Homework

- Compare the resolution of the NCO at high frequency vs. low frequency output
- Find out where the performance of an NCO and a PWM cross
- What is the potential impact of the period jitter on the NCO noise signature? Compare to a traditional PWM.

Online Resources

http://microchip.com/nco - Core Independent Peripherals, NCO Overview
TB3071 - Voltage-Controlled Oscillator with Linear Frequency Output
TB3097 - Digital SMPS - Buck Converter using the NCO Peripheral
TB3102 - Digital SMPS - Boost Converter Using the NCO Peripheral
AN1050 - A Technique to Increase the Frequency Resolution of PWM Modules
AN1470 - Manchester Decoder Using the CLC and NCO
AN1476 - Combining the CLC and NCO to Implement a High Resolution PWM
AN1523 - Sine Wave Generator Using the NCO Module
http://www.sebulli.com/picrx – Radio Receiver using the PCI16F1713 NCO

HLT – Hardware Limit Timer

Core Independent

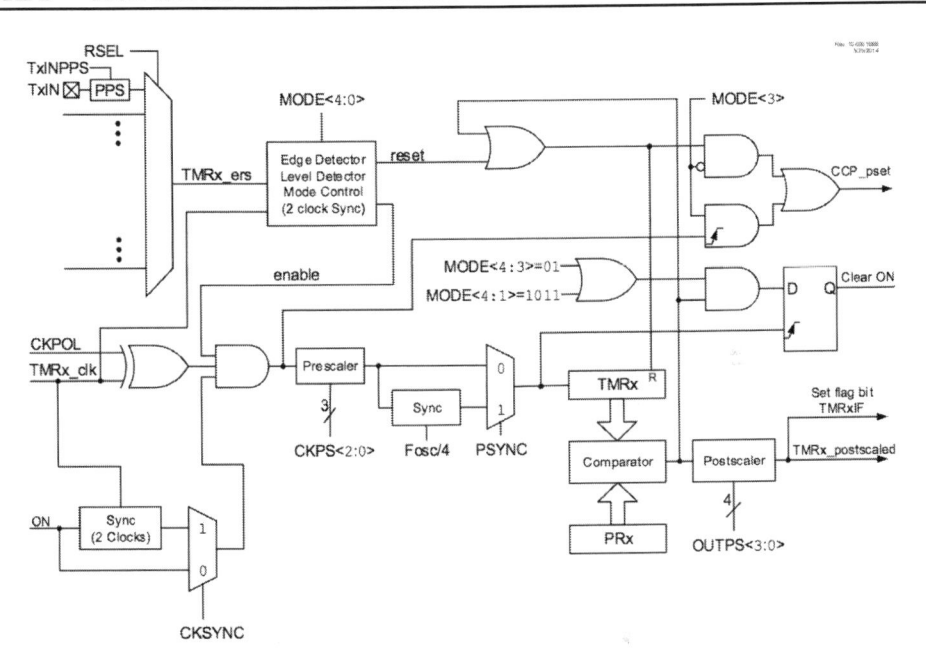

Figure 2.10: 8-bit (Even)Timer with Hardware Limit Block Diagram

Introduction

For all the functionality found in the 8 and 16-bit standard timers used on PIC microcontrollers for the past 25 years it appears that there was one fundamental feature missing: the ability to start/stop and reset the count depending on a hardware external event. Mind these functions could all be achieved in software, with the CPU responding to period events and eventually *manually* clearing the timer, enabling and disabling it. But in the spirit of the Core Independent Peripherals, such operations should not require any CPU intervention and the corresponding control inputs/events should be easily connectable directly with the other peripherals inside the chip. The simplest answer to this need was to develop the Hardware Limit Timer, which is in practice a (much) enhanced 8-bit (Even) timer.

Comparing Figure 2.10 with Figure 2.4 it might not be obvious that the two are really close relatives.

Focus on the bottom part of the block diagram and you will immediately recognize the prescaler block, followed by a timer/comparator/period register structure, in its turn followed by a postscaler.

The new structure on the top of the block diagram can be easily split in two more sections. To the left you will recognize a large Input Signal Multiplexer (ISM) which provides us with a lot of options to get in the *Timer Reset* signal. The right top portion is there to provide further flexibility (modes) to control whether the timer operation has to be periodic or *one-shot*.

How it works

An HLT is first and foremost an 8-bit timer. As such it has the standard selection of pre and post scalers to generate periodic events including participating to the CCP-PWM modules operation.

In addition to the standard 8-bit timers features an HLT provides:

- A much expanded selection of reference clock inputs (typically including more choice of oscillators, CLC outputs, zero cross detect events, and an external pin.

- One-shot operation, that is the ability to automatically stop upon completion of a single period (instead of resetting and starting over immediately).

- Most importantly a *Timer Reset event* input, that can be used to reset the timer count before it reaches the set period/pulse duration.

- A Reset/Stop/Start mode that allows the same Timer Reset event to hold the counter after a reset until a selected edge is detected.

Applications

A practical example can perhaps best illustrate the primary use of the Hardware Limit Timer. Imagine any application where an action is periodically initiated by the microcontroller only to be confirmed by a sensor input after a short delay. It is then possible to imagine a condition where the sensor input is missing or arrives too late. This could indicate a fault or in general an exceptional event that could pose a safety risk and needs to be addressed immediately. A traditional approach would require us to set a timer with the appropriate *time limit*. Upon this timer period expiration the MCU would need to receive an interrupt prompting it to take defensive action. But interrupt response times are by definition un-predictable (if more than one interrupt is enabled at any given time), interrupt latencies can be prohibitively long (tens of microseconds when nanoseconds might be needed) and in general software complexity and its debugging can get out of control as soon as a few asynchronous events are involved. Such an example application could take the shape of a motor control system (with a Hall sensor providing rotational feedback) or a switching regulator (power supply) with a peak current feedback measuring the current surge in the inductor.

The HLT helps us solve these problems with extreme simplicity by activating automatically an output when a timer reset input is not received *before* the timer (limit) period is reached. The HLT output can in its turn be chained directly into other peripheral's inputs such as a PWM auto-shutdown, or output to a pin to drive directly a safety switch. This is really important in the light of the Core Independent Peripheral philosophy: autonomy, fast response, CPU workload reduction!

Limitations

The HLT is after all just an 8-bit timer, so its resolution is somewhat limiting naturally the scope of its application.

When longer time-out periods are required with a 16-bit or higher granularity, you might want to take a look at the Signal Measurement Timer instead.

When very short pulses (as in tens of nanosecond wide) must be produced in response to an external event (re-triggerable one shot) consider using the COG asynchronous dead band control feature instead.

MCC API

The MCC treats the HLT only as a special case of the 8-bit timers. So you will not find it in the list of available peripherals. When activating one of the 8-bit timers (on a devices featuring the HLT) you will simply notice how the corresponding dialog window is much more rich in terms of reference clock options (the clock ISM) and both a Reset Source and Start/Reset Option selection boxes (ISMs) are added to the list.

The generate file will be simply called ***tmr2.c*** (or other even number) and the API will include only a couple of new functions such as:

- TMRx_ModeSet() allowing to change dynamically the HLT modes

- TMRx_ExtResetSourceSet() which allows to change dynamically the timer reset input source.

PinOut

The HLT is found almost exclusively on recent devices with Peripheral Pin Select, so it is possible to assign the Timer Reset input to any available digital pin. The only exception is represented by the PIC16F1612/3 models.

Homework

- Compare the HLT behavior with a WDT module.
- Compare the HLT behavior with the SMT when used in Window Measure mode.
- Compare the use of the HLT (in single shot mode) to a CWG/COG when a short pulse is required.

Online Resources

TB3122 - Hardware Limit Timer on PIC Microcontrollers

SMT – Signal Measurement Timer

Core Independent

Introduction

The signal measurement timer is perhaps the single most intriguing timing module introduced in the PIC16F1 family of core independent peripherals. Although the block diagram (Figure 2.11) might be intimidating at first sight, I would strongly encourage you to take a brief look at the individual modes presented below as I am certain you will find many of them new and uniquely useful in your future applications.

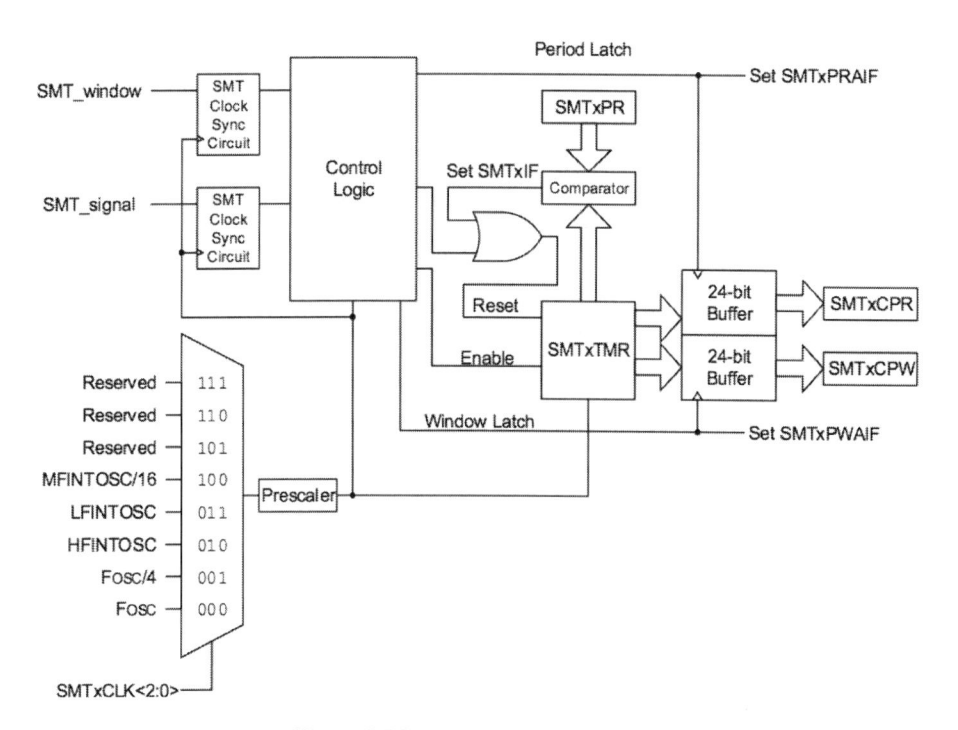

Figure 2.11: Signal Measurement Timer Block Diagram

How it works

The SMT is fundamentally just a 24-bit timer with three input sources (each with its own input signal multiplexer or ISM) and two (24-bit) large output registers, or *measurements*. The timer includes a (24-bit) period register and a comparator that will produce an output *match or timeout* event. The possible permutations of the use of three inputs as *measurement input (signal)*, *gating (window) input* and *reference input (clock)* produce a total of 11 unique modes that we will review shortly in the following to reveal their potential applications.

Period and Duty Cycle Acquisition Mode

In this mode the SMT uses only the input signal and the reference clock to perform a cycle by cycle measurement of the incoming signal period and duty cycle (Ton). The two values are available in the output registers for the entire subsequent period, thanks to a double buffering mechanism.

Note that a similar measurement could be achieved in many alternative ways using traditional timers and interrupt resources (see a complete analysis of such options in Application Note AN1473) but in all such cases either a large amount of CPU cycles were required, RAM and Flash memory was used for creating the required state machine and/or there were severe limitation in accuracy of the measurements and or signal input characteristics.

The SMT removes all such limitation and all timing constraints for the CPU according to the best core independent philosophy, providing a convenient and accurate measurement of the input signal.

Applications of this mode include the cycle by cycle decoding of a PWM incoming signal, which is common in automotive applications, but also in several smoke detectors and fan control applications.

High and Low Time Measurement Mode

This is essentially identical to the previous mode except this time the count is reset on the falling edge. The result is a separate measurement of the Ton and Toff time. All previous considerations and applications apply.

Gated Window Mode – Averaged DC Measurement

In this mode the input signal ON time is measured (gating the clock) over intervals that are controlled by the window input. Once every window period, the new accumulated ON time is captured in one of the two output registers.

The result is a measurement of the input signal Ton *averaged* over the window period. Or in other words an *averaged Duty Cycle* measurement of the input is obtained. This can be useful in applications where the input signal can contain significant amount of noise (or jitter) and rather than an instantaneous value of the duty cycle, an averaged measurement is preferred.

Time of Flight Measurement Mode

In this mode the two *signal* and *window* inputs are used to Start and respectively Stop the count of the reference (clock) input. This is useful in applications that require the precise measurement of the time between two distinct events. In retrospect, now that you have discovered this mode, it makes you wonder could we live without it?! The possibly alternative (before the SMT) was once more the use of Interrupt on Change functions, state machines and in general lots of CPU cycles wasted to achieve what the SMT can do in this mode with absolute simplicity.

Timer Mode

Timer mode is one of the most basic modes. Here only the SMT clock input is used, but we take advantage of the (large) period register and use the match event to produce a long time base. Essentially the behavior is not dissimilar from one of the 8 or 16-bit timers, except that we have 24-bit of dynamic to achieve longer periods or higher resolution.

Window Measure – Big HLT

In the Window Measure mode, the *window* input is used to *capture* the count on each rising edge (the polarity can be inverted if needed) and immediately reset the count. This provides a cycle by cycle measurement of the period of the window input. But since we have still the period register being constantly compared with the running count, we can also obtain a second output event (timeout) should the window input signal period extend for too long.

Another way to look at this is to compare it to the behavior of the a Hardware limit timer module. The window input is essentially the reset event of the HLT and the period register becomes the timeout period. The big advantage is that we have here a 24-bit dynamic range to increase timeout length or to increase the resolution.

Gated Timer Mode

This mode is identical to the 16-bit timers gated modes. The input signal is used to gate the reference clock, but this time with 24-bit of resolution.

Capture Mode

This mode is identical to the use of a 16-bit timer module in combination with a CCP module configured in capture mode. Once more the resolution is higher and the measurement is double buffered.

Asynchronous Counter Modes

There are then three Asynchronous Counter modes (Counter, Gated Counter and Windowed Counter) that do not make use of the reference (clock) input, but simply use the *signal* input to count pulses and eventually the window input to gate the previous similarly to what performed by the 16-bit timers except with a larger dynamic and/or resolution.

MCC API

The MCC handles the SMT similarly to the CCP modules by grouping its modes in three major categories: capture, counter or timer. Each categories corresponds to a different dialog window which allows to complete the configuration eventually achieving all the 11 possible modes presented.

PinOut

The HLT is found almost exclusively on recent devices with Peripheral Pin Select, so that Window, Signal and Clock inputs can be easily assigned to any available digital input. Perhaps the only exception is represented by the PIC12F1612 (8-pin) and PIC16F1613 (14-pin) models.

Homework

- How would you use the SMT if you needed only 8 or 16-bit resolution?
- How would you simulate the SMT Time of Flight mode with a more traditional timer structure and possibly a CLC?

Online Resources

AN1473 - Various Solutions for Calculating a Pulse and Duty Cycle
AN1779 - Sensored Single-Phase BLDC Motor Driver Using PIC16F1613

Chapter 3 - Input / Output

I/O Ports

Enhanced

Introduction

Digital I/Os are perhaps the most basic building block of any embedded application. PIC microcontrollers have been known in the industry for providing the strongest and most reliable I/O structures and modern PIC16F1 devices are no different. I/O strength is one of the greatest benefits an 8-bit architecture has to offer. The small size of the core allows the designers of the PIC to use mature CMOS processes that handle with ease very high currents (up to 100mA, see the High DRV feature later in this chapter) without increasing the cost of the device. But recent generations of PIC microcontrollers have added a lot of flexibility to the design of the I/O ports, in some cases these features increase the application safety, in others it simplifies the application design, reduce external component count and overall cost.
On an 8-bit PIC, I/Os are grouped by 8 in *ports identified by* letters (A, B, ..).

TRIS – Tri-state control

All I/O ports have a direction control register. In the PIC tradition this feature is controlled by the TRISx registers (TRISA, TRISB...) and the interpretation of the value of each bit is quite mnemonic: I – Input, O – Output.
Mind, what this group of register truly does, is to enable or disable the output part of the I/O circuit. A PIC pin after all is ALWAYS available as an input!

PORT – Direct Pin Access

The PORTx group of registers is used to access *directly* the input pin (digital) value. By directly I mean the actual electrical value present on the outside of the I/O driver structure. The distinction is important because, if a strong load is present on the pin, even if we are driving it as an output to the high (1) level, we might be able to read a zero (low). This can be used to advantage to detect such conditions or to detect conflicts when a potential exist for two devices to try to control the same external resource.

Note that PORTx registers can also be written to, with the effect of loading a value on the output latches.

LAT – Output Latches

The LATx registers are featured on all PIC16F1 devices (borrowing from the PIC18 architecture) to provide access exclusively to the output latches.

Note that LATx registers can be read from as well, with the effect of retrieving the last value written into the latches.

LAT registers should be preferred for all writing operations to I/O ports.

ANSEL – Analog Select

Analog and digital functions don't always mix well. While it is possible to read an analog value from a pin (with the ADC) and at the same time to read its digital value (from the corresponding PORT register) this can have unintended consequences. When the input voltage gets close to Vdd/2, both sides of a CMOS input gate are partially biased and a large amount of current can be sinked from the supply rails. Proper design practice is to avoid such conditions by disconnecting the digital input (and grounding it). This is accomplished with the ANSELx registers. Setting one of the bits, configures the corresponding port pin for analog use only.

WPU – Weak Pull Ups

Weak pull up circuits are provided on most I/O pins to save the cost and space of any external resistor. They can be enabled individually (pin by pin) by setting the corresponding bit in the WPUx registers.

ODCON – Open Drain Control New

Most recent families of PIC16F1 microcontrollers (such as the PIC16F161X) offer the option to *simulate* an Open Drain drive by configuring the ODCONx registers. In the past this was often done in software by setting the output latches to zero and toggling the TRIS registers instead. With the advent of the Core Independent Peripherals it is important to have this option in hardware so that the peripherals can drive the outputs directly without CPU intervention. Keep in mind though that this is not a real Open Drain circuit and therefore it cannot be used with voltages above Vdd.

SLRCON – Slew Rate Control

The strength of an I/O drive can be a problem in some cases, where a long trace on a PCB can become an unintended antenna. Most recent models of PIC16F1 do feature a slew rate control register to reduce noise emissions.

INLV – Input Level

Some application require the microcontroller digital inputs to apply a small amount of hysteresis to avoid false or double commutation in the presence of noise. For such cases, selected ports have the ability to switch their inputs from TTL to Schmidt Trigger (ST) mode when the corresponding bit is set in the INLVx registers.

IOC – Interrupt On Change

All pins on all ports can be configured to generate an interrupt when a rising or falling edge is detected. This behavior can be controlled with a pair of registers associated with each port (IOCxP and IOCxN).

Note that any number of pins can be configured to generate an interrupt event. Similarly to all other interrupt sources, the IOC can be used to *wake-up* the processor from Sleep. If the IOC module interrupts are enabled (IOCIE), then an actual interrupt sequence will be started, otherwise the processor will continue executing from the instruction immediately following the Sleep command.

HIDRV – 100mA

Finally, while most/all pins of PIC16F1 microcontrollers have maintained their famous ability to drive (sink and source) up to 20 mA of current, recent models have pushed the envelope to provide up to 100 5A. This is sufficient current to drive even the dullest TRIACs (in the IV quadrant) or perhaps small relays. It is one more reason to love the strength of 8-bit technology.

Once more this is a feature that you will find in the PIC16F161X family and on two selected pins that are by default assigned to the outputs of the COG module (RC4, RC5).

Online Resources

TB3009 - Common 8-Bit PIC® Microcontroller I/O Pin Issues
TB3013 - Using the ESD Parasitic Diodes on Mixed Signal Microcontrollers

TB3061 - Interrupt-on-Change Operation for Mid-Range Microcontrollers
AN1081 - Interfacing a 4x4 Matrix Keypad with an 8-Bit GPIO Expander

CLC – Configurable Logic Cells

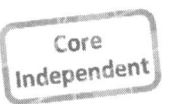

Core Independent

Introduction

The introduction of the Configurable Logic Cell in the PIC16F150x family was perhaps the defining moment in the evolution of the Core Independent Peripheral philosophy. It is with the CLC that the idea of hardware blocks relieving the CPU of part of its workload come first to life. While there had been many previous attempts at defining a cross between a traditional microcontroller and a programmable logic device, the CLC was first to strike the perfect balance. A "puddle of gates" was its original nick-name and in that humble definition lies in my opinion the key to its success.

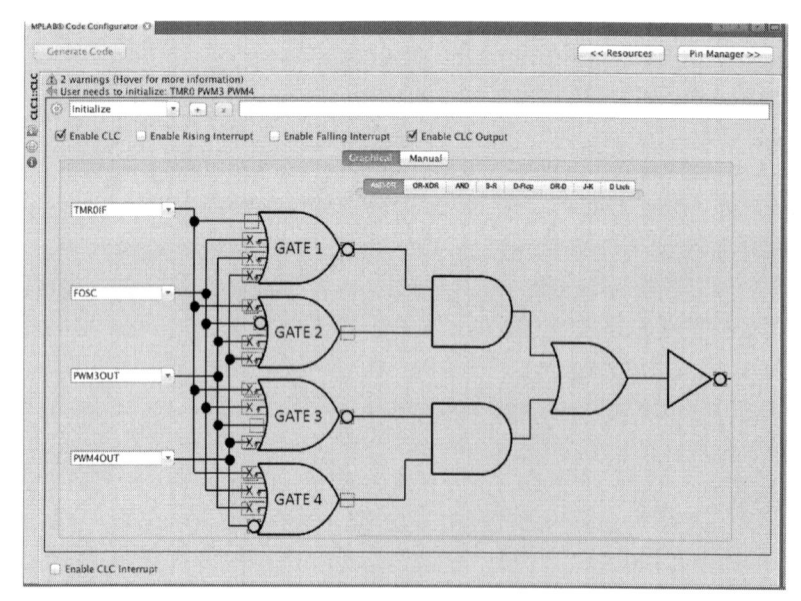

Figure 3.1: MPLAB Code Configurator, CLC Dialog Window

How it works

A Configurable Logic Cell is just that, one single macro-block as if it was just taken from a larger programmable logic array. Up to four of them have been designed into current generations of PIC16F1 microcontrollers with the typical parsimony of he PIC architects. Clearly not enough to design complete new peripherals but plenty to connect existing peripherals (as glue logic inside a chip) and enable small *chains of events*.

Each CLC has the same structure: four input signal multiplexers (ISMs), each input feeding into a combinatorial stage, eventually reaching into a logic function chosen among eight available: AND-OR, OR-XOR, AND, S-R Latch, D-Flop, Or D-Flop, J-K Flop, D-Latch.

The input multiplexers are feeding in signals from the outside world, but most importantly have access to a large number of internal signals, coming from the peripherals surrounding the CLC, timers, clocks, and interrupt events.

The output of each CLC is in its turn made available to the input selectors so that relatively complex state machines can be built. But the same output can be used as a trigger for other peripherals, can generate interrupts and can drive directly an output pin (if required).

There are three important characteristics of the CLC that make it so useful and flexible:

- Each Configurable Logic Cell is completely asynchronous with the rest of the microcontroller hosting it. The device can be put in sleep mode, stopping the core and all oscillators on board, while the CLC continues to function normally.

- CLCs are fast. The propagation time of its logic is in the range of tens of nano seconds. CLCs are also eXtremely Low Power. Each cell, individually enabled adds only an imperceptible amount of power consumption to the device (nano Ampere), too small to be measured.

- CLCs are configured in RAM and therefore their configuration can be changed dynamically during the life of the application.

Applications

From the three properties listed above we can infer some important consequences:

- Being Asynchronous and eXtremely Low Power, the CLC logic is a perfect candidate to implement intelligent wake-up mechanisms. Many low power (battery operated) applications that spend most of their time in sleep and otherwise would require frequent (periodic) wakeup to check the status of sensors and respond to external stimuli, can now be optimized by allowing the CLC to perform a (combinatorial) prescreening of the inputs and waking up the CPU selectively, only when strictly necessary.

- Being fast, the CLC can connect peripherals and drive output directly, even bypassing the CPU (if necessary) to remove application timing constraints (often cause of software complexity) and increase system responsiveness and overall safety.
- Being configured in RAM, the CLCs can be re-used. As the application transitions into different states/modes, so the CLC can be re-configured to assist the CPU in different tasks.

Limitations

The small number of CLCs offered (maximum 4 as of this writing) is helping keep the complexity low but it is also certainly a potential limitation in advanced applications.

The low propagation delay of the logic cells is often mis-understood. While incoming signals propagate extremely fast inside the chip and through the logic, once they reach an output pin, they are in fact subject to the same constrains of all the other I/O pin output drivers, which are typically designed to the maximum device clock (16 or 32MHz). In other words, even if a 10ns propagation delay is achievable, you won't be able to output a 100MHz square wave on a pin.

Further, there is no guarantee that all CLC paths will have the same exact propagation delay. Combine this with the fact that they operate asynchronously and you will soon realize that you have a potential for glitches. This should be no surprise to any serious digital designer, but can be cause of much puzzlement for the novices. In short, the CLCs are like sharp knifes at the table, you must handle them with care!

MCC Support

The MPLAB Code Configurator provides excellent support for the CLC. As seen in Figure 3.4, the CLC dialog window uses a very effective graphical representation of the configurable cells and makes it real easy defining (multiple) initialization functions.

The MCC generates separate source files (clc1.c, clc2.c ...) for each individual CLC module.

PinOut

On the PIC16F150x family, the CLCs are disposed so that each module has a fixed output pin on a different side of the QFN (square) package. This was meant to facilitate using the CLC as a PPS of sorts in devices that were lacking such feature.
Most all new PIC16F1 being designed today do include a PPS feature and with it the freedom to choose any available digital pin as a CLC input or output.

Homework

- The CLC can be used to allow a number of internal signals (and their complex logic combinations) to trigger ADC conversions. Check the ADC module documentation to find out how this can be done.
- By operating asynchronously and independently from the microcontroller core the CLC could be used to identify anomalous states and provide additional checks in safety applications.
- When a CLC input gate is not connected, are its inputs floating?
- How would you allow your application inject a signal into the CLC? (Hint: Look at the input gates output-polarity control)

Online Resources

http://microchip.com/clc – Core Independent Peripherals – CLC Overview
DS41631 - Configurable Logic Cell Tips 'n Tricks
TB3096 - Pulse Code Modulated (PCM) Infrared Remote Control
AN1450 - Delay Block/Debouncer
AN1451 - Glitch-Free Design Using the Configurable Logic Cell(CLC)
AN1470 - Manchester Decoder Using the CLC and NCO
AN1476 - Combining the CLC and NCO to Implement a High Resolution PWM
AN1606 - Using the Configurable Logic Cell (CLC) to Interface a PIC16F1509 and WS2811 LED Driver
AN1660 - A Complete Low-Cost Design and Analysis for Single and
Multi-Phase AC Induction Motors Using an 8-Bit PIC16 microcontroller

PPS – Peripheral Pin Select

Core Independent

Introduction

The Peripheral Pin Select is a feature that was pioneer originally on very large 16-bit and later 32-bit microcontrollers in order to simplify the allocation of the many peripherals available on chip. With time though it became clear that even the smallest PIC16F1 devices were experiencing the same problem: too many peripherals for too few pins!
The PPS solves this problem by allowing the designer a great degree of freedom in mapping digital I/O functions to pins.
Most all new PIC16F1 devices introduced since 2013 have been offering this new feature and this trend is expected to continue and expand in the future.

How it works

The PPS works its magic by introducing two distinct sets of multiplexers:

- Each pin output driver is given a (function) multiplexer so that one (and only one) of a number of internal peripherals can be selected to control it.
- Each peripheral module input is given a (pin) multiplexer so that it can select one (and only one) pin to take its input from.

Notice that this arrangement does naturally prevent conflicts. It is not possible to have multiple peripheral attempting to control the same output, or multiple (input) signals trying to reach the same peripheral module.
The PPS configuration is stored in RAM and therefore must be set at run time during the I/O initialization phase. It is protected by a lock mechanism similar to that used by the Flash Memory and Data EEPROM. A special unlock sequence must be followed exactly in order to modify the PPS control registers. Further, a device configuration bit can be set (in Flash) to make the lock permanent, or rather *one-shot* until the device reset (POR/BOR/MCLR).

Applications

The PPS is of great help in optimizing the use of internal resources of a PIC.
It can also help reduce the complexity of a PCB and possibly eliminate some of the contortions involved in routing signals around a board, removing/reducing the use of Vias and perhaps removing the need for additional layers saving cost.

Optimizing the pinout means, among other things, having the ability to control the placement of "noisy" (high speed commutation) signals away from sensitive (high impedance analog) inputs, reducing parasitics and matching trace lengths.

The PPS can also be used to share resources across multiple pins, such is the case for example when a serial interface can be shared between two ports that are used alternatively. The same UART can be presented alternatively to a printer connector during certain application phases or on a separate connector to drive a display or a debugging console at other times.

Lastly, it is possible to use the PPS to present the same output function on multiple pins effectively multiplying the fan-out by connecting together multiple CMOS output drivers, which will naturally and gracefully share the load (within the absolute limits of the package/device).

Limitations

The PPS size and complexity could grow exponentially as the number of pins and peripherals available on a chip grow. To keep the cost in check the 8-bit PIC architects typically limit the PPS to groups of 20 pin maximum. This means that on small pin count devices (8/14/20-pin) all digital pins are available for all functions enabled. But on 28 and 40 pin devices, the pin out is pretty much split in two groups. There is no such a thing as unlimited power!

When routing the outputs of the I^{2}C interface (MSSP module), care must be taken when selecting possible pin configuration for the SDA and SCL functions. The original bus specifications dictate the use of special I/O drivers that differ from standard CMOS port specifications and are only available on two special pins. Most applications, such as serial EEPROMs, small sensors etc. will typically not be affected and will allow you instead to make a more liberal selection take full advantage of the PPS routing capabilities.

Only digital peripherals/inputs can be routed by the PPS. Analog peripherals require a different set of multiplexers and are in general much more limited in their choice to keep noise and cost in check.

MCC API

The MPLAB Code Configurator does a great job of tracking the use of the PPS via the Pin Manager window.

Once the selection is done, the MCC adds the appropriate PPS commands (and locks it) into the `PIN_MANAGER_Initialize()` function found inside the `pin_manager.c` source file.

PinOut

Nothing speaks of the power of the PPS like the view offered by the MCC Pin Manager in the example of Figure 3.2. Here a PIC16F1708 user had selected the COG module and was presented with unlimited choice of where to allocate its input and output pins on any/all but the power supply pins.

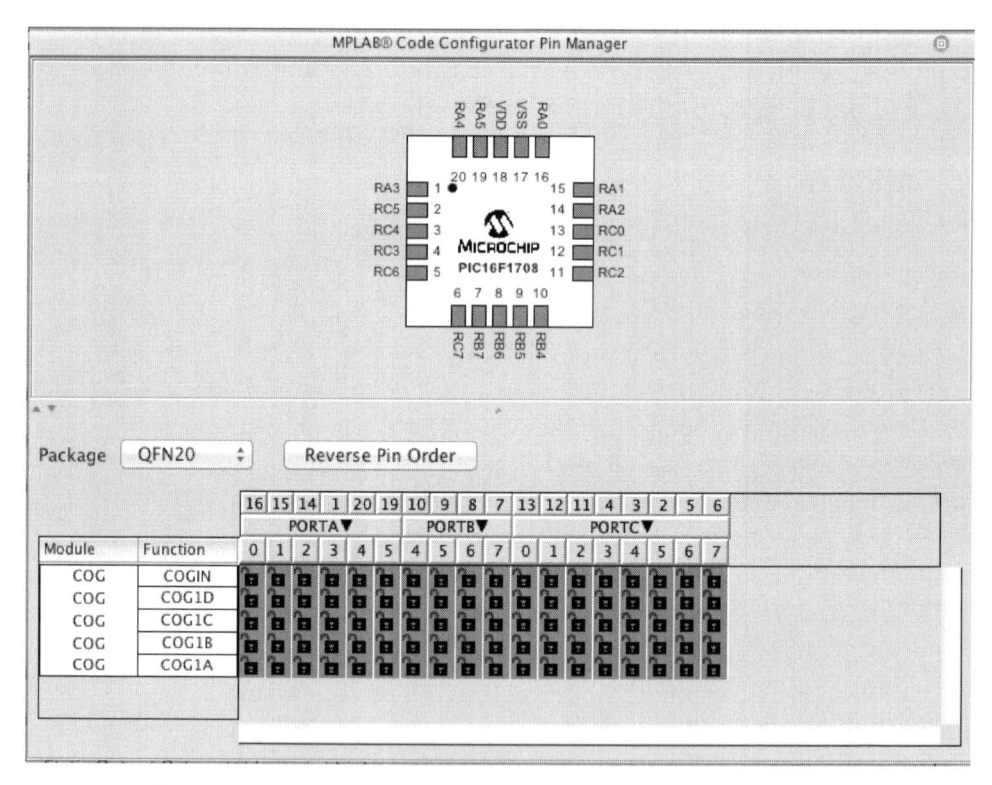

Figure 3.2: MCC Pin Manager view of the PIC16F1708

Homework

- Can I/O Port pins be re-mapped?
- Which peripherals can use the PPS and which ones don't?

- How about PWM output steering and the PPS? Double steering?
- Can the ZCD input pin be re-mapped?
- What would limit the maximum fan-out (drive current) if you could parallel many/all pins as output for a single peripheral?

Online Resources

http://microchip.com/PPS – Core Independent Peripherals – PPS Overview
TB3096 - Pulse Code Modulated (PCM) Infrared Remote Control
TB3098 - PIC16F170X Peripheral Pin Select (PPS) Technical Brief

DSM – Data Signal Modulator

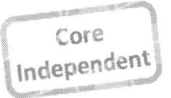

Introduction

The data signal modulator is essentially a digital mixer. A modulator input signal is used to switch the output between two different sources (carriers). When the modulator signal is high, the high-carrier input is connected to the output. When low, the low-carrier input is connected instead. The carrier signals can be obtained internally (from PWM modules and/or system clock) or from external pins. The modulator signal can also be generated internally but by a larger number of sources including comparator outputs, asynchronous sources (UART TX) and synchronous serial ports I^{2}C and SPI, or an external pin. Optionally the modulator signal can be synchronize with the carriers to avoid output glitches.

The flexibility of this mechanism allows for a number of modulation schemes among which: FSK, PSK and OOK (when one of the two carriers is grounded).

How it works

As illustrated in Figure 3.3, the DSM is essentially performing the AND of the modulator input with the two carriers. The three input signal multiplexers (ISMs) dominate the module design which is otherwise quite small and simple.

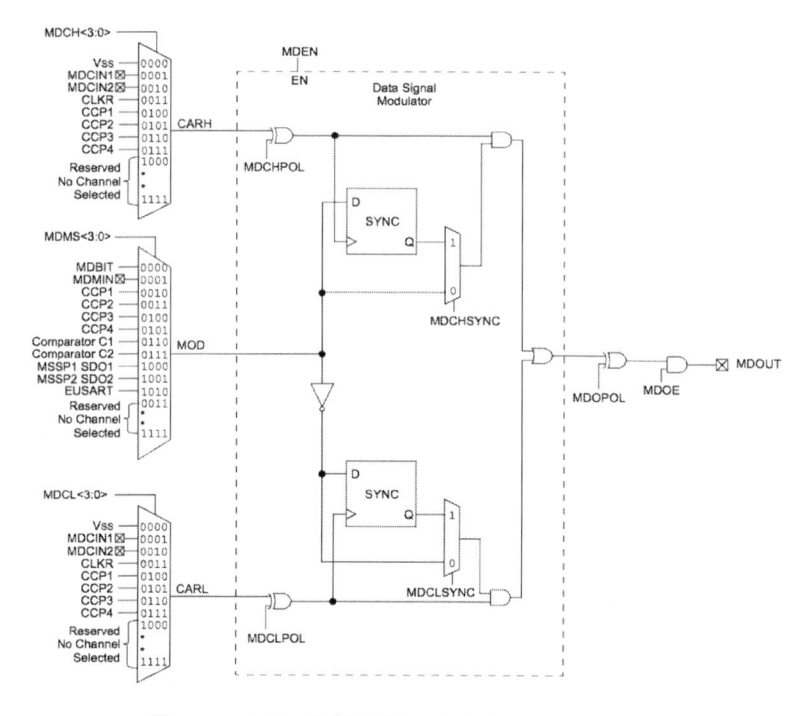

Figure 3.3 - DSM Block Diagram

Applications

Example of DSM applications include: infrared modulation of the asynchronous serial port, dimming of solid state lighting (High Brightness LED) devices, and radio signals modulation.

Limitations

DSM uses are essentially limited by the maximum speed of the I/O output drivers of the specific device. A 32MHz capable PIC16F1 will have I/O drivers optimized for an 8MHz output square wave, representing the highest possible carrier that can be used. Higher frequency clock sources are likely to produce a distorted output signal.

MCC API

The MCC allows to select the desired configuration and generates a default DSM_Initialize() function plus a number of smaller (singe-liners) functions to manually control the DSM operation, start and stop modulation and control the output pin.

PinOut

The DSM was originally offered on devices without a PPS and therefore limited its input source pins to only three possible options. More recent versions add the PPS flexibility to expand the number of options to any available digital I/O.

Homework

- How is a DSM different from a CLC AND-OR block?
- Could you design your own DSM perhaps using two or more CLC blocks?

Online Resources

TB3126 - PIC16(L)F183XX Data Signal Modulator (DSM) Technical Brief

ZCD – Zero Cross Detect   New

Introduction

For many years it has been an accepted practice to make use of the (ESD) protection structures (often represented as diodes) found on every I/O pin to detect the presence of a high voltage on a given I/O by simply inserting a large series limiting resistor.

While this certainly works, and no damage whatsoever is produced even if thousands of volts are presented at the extremity, there are undesirable side effects. When AC and in particular negative voltages are applied, the current flowing through the resistor is routed by the protection "diodes" through the device substrate and eventually to the ground (Vss pin). The effect can be as innocuous as an offset added to the ADC reading, an inaccuracy of a voltage reference (FVR), but could also mean disrupting a low power oscillator or triggering a device BOR.

These effects happen to be just more noticeable today when so much more (sensitive) analog circuitry is present on even the smallest PIC microcontrollers and when all internal circuits have optimized for the lowest power consumption.

The Zero Cross Detect module was designed to avoid all such dangers by providing means to keep the input pin always perfectly within the nominal voltage range and eliminating completely any substrate injection current.

How it works

The new Zero Cross Detection circuit is based on the idea of maintaining the input pin not just *within* the normal voltage range, but *precisely* at one specific voltage. This is accomplished (as illustrated in the block diagram of Figure 3.4) by comparing the input voltage against an internal reference (see Chapter 7-FVR) and then driving alternately one of two current sources.

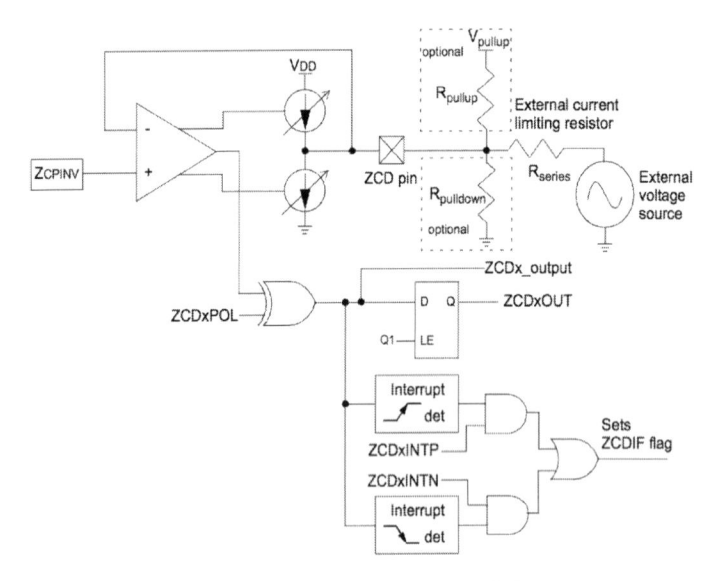

Figure 3.4: Zero Cross Detect Block Diagram

When the external voltage source pulls up one end of the series resistor, the low side current source will be activated to draw current into the device and produce a voltage drop across the series resistor countering its effect. Similarly, when the external voltage source is pulling low one end of the series resistor, the high side current source is activated, sourcing current out of the device so to produce an opposite voltage drop. When the series resistor is properly dimensioned (~1MOhm) a relatively small amount of current (<0.3 mA) is sufficient to counter typical AC line voltages (~240V).

The problem of detecting the zero crossing event is therefore reduced to identifying the moment when the two current sources are being switched.

Applications

Zero cross detection is commonly required in applications that make use of AC power possibly in combination with the use of TRIACs or IGBTs to perform control via phase cutting techniques.

But ZCD can also be used simply to detect the line frequency as a (relatively) stable timing reference.

In some small appliances, ZCD has been used simply to detect the presence of a phase when implementing inexpensive safety switches and in order to reduce wiring cost.

Limitations

There is a small phase error caused by the offset introduced by the comparator as we are using a reference voltage of ~1V rather than 0V, but this can be easily accounted for.

As of this writing, there are no PIC models with more than one ZCD module, although plans to introduce multiple such inputs in future product families have been debated.

MCC API

Beside the customary ZCD_Initialize() function, the MCC offers only a single polling function ZCD_IsLogicLevel(). If interrupts are enabled, they are automatically inserted in the Interrupt Manager and a ZCD_ISR() template is added to the zcd.c source file.

PinOut

The ZCD is essentially an analog circuit and as such it cannot be easily re-routed by the PPS. In all current implementations the ZCD is available only as fix pin option.

Homework

- Would the ZCD detect the zero crossing point if used after a rectifying bridge?
- What is the impact of power line communication signals (X10) on the ZCD and would you filter it out?

- What are the trade offs when using ZCD to detect high(er) frequency AC inputs?

Online Resources

http://microchip.com/ZCD – Core Independent Peripherals – ZCD Overview
TB3013 - Using the ESD Parasitic Diodes on Mixed Signal Microcontrollers
TB3099 - Extending Relay Life by Switching at Zero Cross

Chapter 4 - Non Volatile Memory

Introduction
Flash memory is the technology of choice for the vast majority of microcontroller applications nowadays. But not all Flash technologies are made equal. PIC16F1 devices, thanks to their minuscule 8-bit core can use a mature CMOS process that can withstand up to millions of erase/write cycles of endurance and guarantees retention for 40 years.
PIC16F1 devices all employ three different strategies to provide the best

Data EEPROM

Electrically Erasable Programmable Read Only Memory, or EEPROM, is an acronym nobody ever spells out in full anymore, probably because there is so much history packed in it that most young programmers don't even know of.
EEPROM is the goto technology for *non volatile* storage of data that can change frequently and for this reason, the memory cell design is focused on maximum endurance, retention and also ease of use.
Data EEPROMs allow individual byte access and automate the data write sequence by performing an erase cycle transparently to the user. The memory array is separate from the main Flash program memory array of the device and therefore access to the array can be instantaneous when readying and slow (2ms typ.) but asynchronous during writing. It is left to the user to either wait idling in a loop for the write sequence to complete or to perform other background tasks (not involving the EEPROM).
All PIC16F1 models featuring a data EEPROM array can guarantee a minimum of 100K erase/write cycles.

XC8 API
The XC8 compiler provides a default small library for access to the EEPROM data composed of the following functions:
- eeprom_write() and eeprom_read() access a single byte at a time.

- eecpymem() and memcpyee() transfer a block of data at a time.

Example

```
/*
 * Project:    EEPROM
 * Device:     PIC16F1829
 */
#include <xc.h>
#include <stdint.h>

// initialize EE at programming
__EEPROM_DATA( 0xA, 0xB, 0xC, 0xD, 0xE, 0xF, 0x8, 0x9);

void main(void) {
    uint8_t     data;
    uint8_t     buffer[8];

    data = eeprom_read( 1);     // read a byte from EE, data=0xB
    data = EEPROM_READ( 2);     // same using the macro version, data=0xC

    eeprom_write( 1, 0x55);     // write a byte to the EE
    EEPROM_WRITE( 2, 0x56);     // same using the macro version

    eecpymem( buffer, 0, sizeof( buffer));   // copy EE to ram buffer
    buffer[0] = 0x00;
    memcpyee( 0, buffer, sizeof( buffer));   // copy data back to EE
}
```

Flash Memory

The Flash program memory array of the microcontroller is designed with different objectives in mind. It must provide long retention for reliable operation over many decades, it must be re-writable with a large number of erase/write cycles, but it must also be very compact and economical as this is by far the largest structure on chip for most if not all the PIC16F1 microcontroller models.

Flash Program memory is therefore organized so that the erase process can be performed by blocks rather than individual bytes. Writing to flash memory is similarly organized so that entire blocks (often referred to as *rows*) of 16 or 32-bytes can be updated at once. This makes the cells packing in large arrays more efficient, and provides for a faster write process when programming the device in circuit.

Reading from Flash program memory is fast and performed in words (14 bit). One word can be read every four clock cycles at maximum device speed.

Writing or better updating the memory contents though require a little more care. First, an erase sequence must be performed on the desired row, then a set of latches must be pre-loaded with the new information (16 or 32 latches each containing a word), then a write sequence can be initiated.

During the erase and write sequences (2ms typical) the flash memory array cannot be used by the core for instruction fetching, therefore the core execution is stalled automatically.

Notice that since the writing process is performed in rows of 16 or 32 words each, the effective writing speed (bytes/s), when compared to a data EEPROM, is multiplied by an equal factor.

All PIC16F1 models can guarantee a minimum of 10K erase/write cycles on the entire contents of the Flash memory array over the entire temperature range (-40 to + 85°C).

MCC API

When the Flash memory module is selected in the MCC dialog window, a standard set of four functions is added to the *memory.c* file, including:

- FLASH_EraseBlock(), FLASH_WriteBlock() two primitives that operate on entire row of words. Note that the *Write* function calls already the *Erase* function before executing the *write command* proper.
- FLASH_ReadWord(), which can read any single word from Flash or HEF memory
- FLASH_WriteWord(), which truly uses the three functions above to read the current content of a row into a buffer of appropriate size (which must be passed as the second parameter), erase the entire row contents, patch the *new* word into the buffer and eventually write it back to memory.

HEF – High Endurance Flash

The High Endurance Flash block is a recent innovation typically offered in models that do not have a dedicated data EEPROM array. The HEF design goal was to achieve the same level of endurance as offered by the data EEPROM, while maintaining the economy and the writing speed of the main flash memory array. For this reason the HEF is currently implemented as a subset of the main program memory array, typically the last four or eight rows.

The result is an array that guarantees 100K erase/write cycles (over a slightly restricted temperature range, 0-60°C) just as a data EEPROM, but reads and writes at the speed of flash memory, 2ms will suffice to write up to 32 bytes of data at once. This can be a great feature for applications that use non volatile memory to store data/application state when a low battery or brown out condition is detected. The faster data can be written (32x faster in this case) and the smaller the capacitors on the system power supply will need to be to keep the application powered during the backup phase.

Since the memory array is shared with the program memory space, the user is given the flexibility to choose if and how much of the HEF space to use for data or for program as required by the application. For the same reason though, during the HEF write time, the core is stalled automatically just as when writing to Flash program memory.

Homework

- When comparing the benefit of using HEF for application state/data backup during a brown-out event, try to visualize the difference in size and cost between a 1 uF capacitor (ceramic, 1x 1x 2 mm) and a 32 uF capacitor (Tantalum/Electrolytic, 5 x 5 x 10 mm or larger).
- Compare NVM endurance (e/w cycles) against the frequency of update of parameters in an application and the application life?
- Vice versa, assuming a 10 years application life, how often can an application save data/state per day when using a High-Endurance Flash block? What is the alternative cost of an external serial Flash and the added complexity/time of accessing data on it?

Example

Application Note AN1673 offers an alternative API to make HEF memory blocks management easier by numbering them (0..3) instead of requiring use of absolute memory addresses. Here is an example of its use.

```c
/* Project:    HEF
 * Device:     PIC16F1509
 */
#include "system.h"
#include "HEFlash.h"

void main(void)
{
    uint8_t r;
    typedef struct {uint16_t ID; char Name[20]; uint32_t Amount;} Record;

    // a block of data that needs saving -- fast!
    Record data = { 0x1234, "HE-FLASH", 42};

    // write data to HEF block-1  (2ms!)
    r = HEFLASH_writeBlock( 1, (void*)&data, sizeof( data));

    // empty the buffer
    memset( &data, 0, sizeof( data));

    // read back its contents
    r = HEFLASH_readBlock( (void*)&data, 1, sizeof( data));

    // read a single byte from block-1 at offset 5
    r = HEFLASH_readByte( 1, 5);

    while( 1);
}
```

Online Resources

http://microchip.com/hef – Core Independent Peripherals – HEF Overview
AN1673 - Using the PIC16F1 High-Endurance Flash (HEF) Block
TB016 - How to Implement ICSP Using PIC16 FLASH MCUs
TB072 - FLASH Memory Technology: Considerations for Application Design
AN1019 - EEPROM Endurance Tutorial
AN1188 - Interfacing with UNI/O® Bus-Compatible Serial EEPROMs
AN1449 - High-Reliability and High-Frequency EEPROM Counter

Chapter 5 - Safety Functions

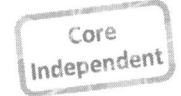

Core Independent

CRC – Cyclic Redundancy Check with Memory Scanner

Introduction

A cyclic redundancy check is an error detecting code that is commonly used in embedded control to ensure the integrity of communication and often the integrity of the microcontroller's own memory (flash) contents. There are many kinds of CRC algorithms generally grouped by the size of the *polynomial* used that translates in its ability to detect a wider range of errors. CRC algorithms are relatively simple to implement in software on a microcontroller as they can be translated in a series of shift and XOR operations, but with the growing size of the polynomial and depending on the amount of data to be checked, the performance requirements on the CPU can get prohibitively large quickly.

In particular, in *safety critical* applications (such as those that require EN/IEC 60335-1, EN/IEC60730 Class B certification and/or UL98 certification), it is necessary to verify the CRC of the entire flash memory area containing the application before first execution, and later, the test must be repeated periodically during the application life.

It is in these cases that a Core Independent CRC peripheral becomes an invaluable tool to reduce the CPU workload and to simplify the application development and testing.

How it works

The CRC module recently introduced on the PIC16F1 family is a hardware implementation of the CRC algorithm similar to that offered previously on larger 16-bit architectures (PIC24/dsPIC). The peripheral can be configured to apply any standard polynomial up to 16-bit, covering a very large number of telecommunication and safety use cases.

After configuration, the CPU can pass data to be checked to the CRC module as it arrives from a serial communication interface, or as a packet of data is prepared for transmission in a RAM buffer.

But the true unique new feature added to this generation of products is the ability to feed automatically from a *Memory Scanner* unit. This allows us to compute the CRC over arbitrary segments (or the entire content) of the device program (flash) memory in the background without any CPU intervention.

Depending on the application, the memory scanner can be configured to operate in one of the following modes:

- *Burst mode,* meant to be used when the CRC calculation has to be performed at maximum speed. The scanner then takes absolute priority and *stalls* the CPU preventing it from tending at any other task while the CRC is being performed. The result is a CRC calculation performed orders of magnitude faster than possible with a pure software implementation.

- *Concurrent mode*, designed for high CRC throughput as well, but instead of stalling the CPU, it allows it to continue operating in between CRC access cycles. The result is a very fast CRC computation but, although the CPU is slowed down, it has still the ability to run some lower priority activities in the background (respond to interrupts etc.)

- *Trigger mode*, is designed to offer a flexible compromise between CRC performance and impact on the CPU performance. A user configurable trigger (generated periodically by a timer for example) defines the rate at which the memory scanner is allowed to steal cycles from the CPU.

- *Peek mode,* is the most gentle of them all. The CPU takes maximum priority and the CRC memory scanner is allowed to use a memory access cycle only when the CPU doesn't, as is the case when a branch instruction is executed (any loop, interrupt ..). This means that the CRC execution speed becomes somewhat dependent on the application code being executed, but means also that the CRC memory scanner becomes completely *transparent* to the application.

Considering how often embedded control application have stringent requirements for interrupt response latency and execution, an additional option (*Interrupt Mode*) is offered by the memory scanner module, so that the scanning can be suspended when executing an ISR, resuming only upon return.

Applications

As mentioned in the introduction, the use of the CRC is often associated with (serial) communication protocols, see LIN bus and 1-wire for typical examples. In *safety critical* applications, industry standards often dictate the use of the CRC module and can benefit from the flexibility of the memory scanner. For example *Burst* mode might be used during the application power-up phase, to speed up the first complete memory check. Later, Trigger or *Peek* mode might be used to satisfy the requirements for a periodic re-check while providing the CPU with maximum performance.

Limitations

The CRC memory scanner operation is independent of the Watchdog. This means that when a high priority CRC mode is selected (burst or concurrent) care must be taken to ensure that the lost (stolen) CPU cycles, do not end up exceeding the WDT limit.
Communication protocols that require CRC polynomials larger than 17-bit cannot be supported. Fortunately these are uncommon in the typical application space of 8-bit microcontrollers.

MCC API

The MCC configuration window for the CRC is quite helpful. It offers a preselection of common CRC polynomials and includes even a test button to verify the peripheral (simulated) operation against test inputs and the expected results for the chosen configuration/polynomial, all without compiling the application or firing an in circuit emulator.
The API includes a few handy functions to load, start, stop and fetch the results off the CRC module.

Homework

- Check out the wikipedia entry for: **Cyclic_redundancy_check**
- See how many uses there are for the CRC algorithm and the polynomials used. How many of those can be covered by the CRC module?

Online Resources

AN1817 - Using a Hardware or Software CRC in Class B Applications

AN1148 - Cyclic Redundancy Code (CRC)
AN1229 - Class B Safety Software Library for PIC MCUs and dsPIC DSCs

WDT – WatchDog Timers

Enhanced

Introduction

A good Watchdog module has been one of the strong features of the PIC since the very first model (PIC16C54) back in 1989. If the enabling configuration bits (OTP) were set, an internal oscillator was dedicated to generate an *independent* clock for the module. This was a great safety feature, and a distinctive one, at a time when most competing devices simply used the system clock (eventually scaling it) to achieve the desired timeout period and the configuration bit was in RAM and therefore easily overwritten. Modern PIC16F1 devices have kept the safety features and expanded on them in several ways to make the Watchdog circuit a very flexible and useful tool.

How it works

Essentially the Watchdog is an independent timer that is supposed to be periodically reset by the CPU (using a special instruction code) when things are *running smoothly*. Should the CPU get *stuck,* whether that is due to an unforeseen *soft* condition (due to poor software design) or to a cosmic ray flipping the wrong bit of RAM (yes these things can happen) it does not matter, the WDT does timeout and cause a complete device reset.

Among the new features added to the PIC16F1 families of microcontrollers, it is worth mentioning:

- Timeout periods ranging from 1 ms to 256 seconds can be configured.
- *Wake up from Sleep,* when a WDT timeout is generated while the CPU is in a low power mode (Sleep), instead of a reset, the CPU is simply awakened with a status bit flagging the particular situation.
- *New* — *Windowed mode,* (on selected models) to increase further the safety of the application (and to prevent *cheating*), a valid reset window can be defined. If the WDT is cleared too early, or all the time in every loop, its usefulness is clearly reduced. The windowed feature helps prevent this.

Safety Functions - 87

- *Soft enable/disable*, for applications that cannot afford the power consumption (no matter how small) of the independent oscillator, the WDT can be configured to (using one of the device configuration bit in flash) be enabled/disabled under software control.
- *Automatic Sleep disable*, makes the WDT automatically disabled when the device goes into a low power mode (Sleep).
- On selected models, it is possible to select an alternate clock input, as a divided value of the MFIntOSC to help ensure that CPU and WDT are never running off the same oscillator.

Applications

Beside the safety feature for which it was designed, the WDT is very often used as a low power period wakeup mechanism.

Limitations

Since the WDT *can* operate even when the CPU is in a low power (Sleep) mode, it has become a precious resource for ensuring periodic wakeup in extreme low power applications. All new PIC16F1 watchdog circuits have sensibly increased the stability and accuracy of the dedicated oscillator (down to ~10%) compared to the historical models were this was famously uncalibrated and wildly inaccurate.

The PIC16F1 WDT is also much less dependent on temperature, which eliminates one of the applications for which it had been famously (mis-)used in the past. For such applications refer to the Temperature Indicator module presented in Chapter 7. (Analog Functions).

Yet the WDT circuit cannot be trusted to produce a precise enough timing as required in certain wireless protocols. In such cases, rely on Idle modes instead and control the power consumption of the device by use of the new Power Down control (where available) and selecting a suitably low power *calibrated* oscillator.

MCC API

The MCC allows the WDT module configuration via the Configuration Bit selection in the System module. The resulting code (pragmas) is placed in the mcc.c source file.

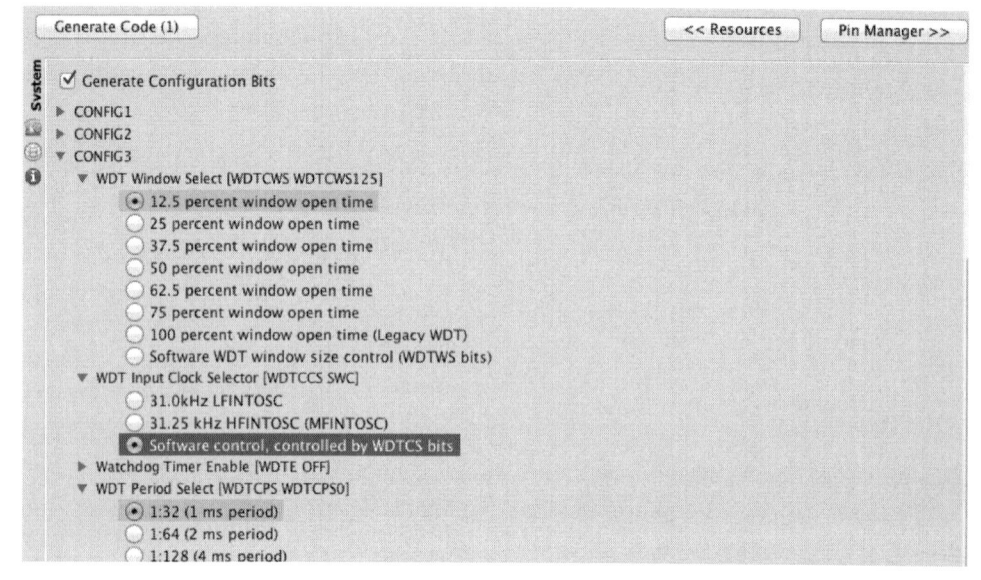

Figure 5.1: System Configuration Bits Setting with MCC

Homework

- Look up the Windowed Watchdog Timer (WWDT) chapter in the device datasheet (i.e. PIC16F161X)

- Notice that devices featuring the WWDT, can now inspect in real time the value of the watchdog timer (via the WDTTMR register) and verify that it is incrementing correctly. This simplifies Class B applications certification/test.

Online Resources

TB3123 - Windowed Watchdog Timer on PIC Microcontrollers

Reset Circuits

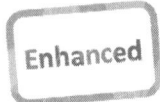

Introduction

Most 8-bit embedded applications are designed to operate in rough environments. Which means unregulated power supplies, noise and in general

Safety Functions - 89

the aggressive interpretation of all the device maximum rating specifications. Anything, in order to save cost (and space) in high volume applications. So even the most basic form of power supply supervision is a luxury that rarely can be afforded. A PIC is expected to survive to anything and always gracefully return back to a functional state. It is for this reason that PIC16F1 devices have several reset circuits that are ready to intervene in the most difficult situations.

POR – Power On Reset

The Power On Reset circuit is the most critical of the whole bunch. It must start operating before any other analog or digital feature of the microcontroller is enabled and therefore it cannot count on any configuration, logic, or advanced analog reference.

The POR of the PIC16F1 series is designed to detect a rising Vdd voltage of appropriate slope (not too slow please) and secure an initial state for the rest of the circuit as soon as a fixed threshold (1.6 V typ) is passed.

Should the supply voltage fall (bounce) after reaching the safe threshold but before the rest of the microcontroller is safely booted (1.8 V), it must rearm safely given a minimum of hysteresis (must descend below 0.8 V typ).

BOR – Brown Out Reset

The Brown Out Detect circuit comes into play *after* the POR has *booted* the device and takes care of fluctuations of the power supply that would take the microcontroller below the nominal safe operating voltage (1.8 V).

BOR thresholds options can be selected using configuration bits and use accurate internal voltage reference resources (FVR) that the POR could not rely on. The BOR circuit has also access to a low power oscillator (LPIntOsc) which allows for a timed delay (64 ms typ) before re-arming when the voltage supply bounces back into the safe zone and before the reset condition is released.

The BOR circuit can be configured to disable automatically when the device enters a low power mode (Sleep) to reduce the power consumption (due mostly to the voltage reference circuit and the low power timer mentioned above).

LPBOR – Low Power BOR   *New*

Selected PIC16F1 devices offer an additional Low Power BOR circuit that is specifically designed for eXtreme Low Power applications. This provides a wider set of tolerances but in returns provides a greatly reduced power consumption.

MCLR – Master Clear

When all the above internal reset circuits are operational, the actual reset input pin (historically referred to as the Master Clear or MCLR in Microchip literature) can be disabled (via a Configuration Bit) and re-assigned to becomes a standard input port pin (RA3).

Since, regardless of its use as a reset, it always maintains a role as one of the *gates* to in circuit programming mode (Vpp), and as such it must be able to detect and withstand a high voltage (~9V) applied to it, its I/O driver cannot be a full push-pull. On most devices this pin becomes an input-only function, while on selected devices (PIC16F155x family) it becomes also available in an Open Drain configuration as the SDA pin of the I^{2}C interface.

FSCM – Fail Safe Clock Monitor

The Fail Safe Clock Monitor is designed to detect the failure of external oscillators. Those are the most susceptible to parasitic effects due to via, PCB layout issues or simply bad soldering joints and EMI.

The module is enabled using one of the Configuration Bits of the device and it works by comparing the selected external clock signal with the internal low power oscillator (LFIntOSC). Should the external oscillator fail to produce at least one commutation during a sampling period (~2ms) a clock failure event is generated. As an immediate consequence, the system clock is switched to an internal oscillator (according to the configuration set in the OSCCON register) that will allow the application to continue to work although with less accuracy. If the FSCM interrupt is enabled, this will also be triggered to allow the application to take the necessary precautions or attempt a new restart.

Online Resources

TB087 - Using Voltage Supervisors with PICmicro MCU Systems
TB3121 - Conducted and Radiated Emissions on 8-Bit Microcontrollers

Chapter 6 - Communication Functions

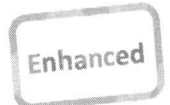

I²C – Inter Integrated Circuit Bus

Enhanced

Introduction

I do remember a time when "real" programmers did not ask for an I²C peripheral only to talk to a serial EEPROM, but maybe I am just too old. Back then I remember we were circulating an example code that in only 40 lines (of assembly of course) implemented a rough I²C master and could read and write data from a 24LC16. To be fair though, in those days the expectations for an embedded application were set much lower. Today we have to support much more complex tasks and bit banging a low level serial communication protocol like the I²C would be a huge waste of the microcontroller potential.

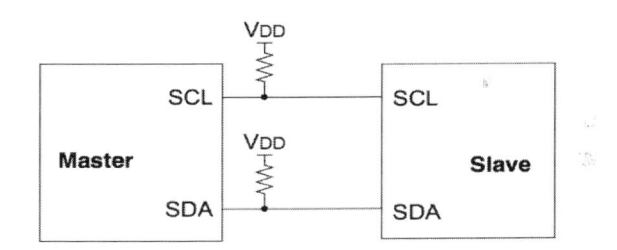

Figure 6.1: MSSP Block Diagram I2C mode

How it works

All PIC16F1 microcontrollers that support the I²C bus do so via the Master and Slave Synchronous Port (or MSSP) peripheral. This is shared with the SPI interface (see next section) to save on configuration registers and buffers in typical Microchip frugal approach.

As the name implies, both Master and Slave functions are supported as well 7 and 10-bit addressing modes to satisfy all common use cases. The MSSP peripheral supports a list of additional functions that are assisting, among other things, in the implementation of SMBus and PMBus protocols (built on top of I²C). These include:

- Address masking, so that a device can be made to respond to any address within a range (vectoring)
- Collision detection, to flag cases where multiple masters are attempting to control the bus simultaneously
- Clock stretching, to provide more time for a slave device to prepare a reply (ack/nack)
- General call detection, to allow a Slave to respond to a call at address 0x00 regardless of the selected address and mask
- Interrupt generation can be extended to each start/stop event to provide maximum control on the communication sequence
- A baud rate generator is used to produces the desired data transfer speeds by dividing the main system clock

Applications

The use of I²C interfaces is typically restricted to the communication of devices that are located in close proximity, on the same PCB. This means usually sensors, displays, non volatile memory devices, but also communication with other microcontrollers located in the vicinity. The synchronous nature of the serial port means that it is not necessary to use high accuracy (crystal) oscillators reducing the cost of the application. Because the I²C interface requires only two wires/pins to establish bi-directional communication, it is often preferred over SPI when space is at premium and the transfer speed (up to 1 Mb/s) is sufficient for the application.

Limitations

The I²C interface can operate while the device is in sleep, in fact when configured in Slave mode, it will wake up the device upon detecting a start condition (if the corresponding interrupt event is enabled) or an address/data byte is received. But the I²C interface still requires a bit of hand holding to complete an entire data transaction, so we cannot truly claim core independent operation (yet).

Since several registers are shared between the SPI and I²C function, once the MSSP peripheral is used for one of the two functions, it is not possible to use the other. So if you need both I²C and SPI interfaces in an application, you will be looking for devices with two independent MSSP peripherals.

MCC API

The MCC differentiates between Master and Slave configurations of the I²C interface but in both cases assumes the use of interrupts to handle complete transactions via an interrupt based state machine.

When the Master I²C mode is selected, the MCC generates a state machine that can execute automatically complex sequences of *transaction requests*. Each transaction request can describe a piece of a typical I²C command, such as sending an address, sending data or reading data.
Transaction requests (TRBs) can be generated using a set of helper `..Build()` functions and queued up using the `I2CMasterTRBInsert()` function.
The state machine runs continuously through the queue and reports the execution status for each transaction and eventually the completion of all tasks in the queue.

When the Slave I²C mode is selected, the MCC generates a different state machine that can emulate the behavior of a typical slave device. The I²C protocol processing is cleanly separated from the device behavior which is defined in a callback function (`I2C_StatusCallBack`). An example of one such callback function is provided in the generated file demonstrating emulation of a serial EEPROM using a RAM buffer.

PinOut

The I²C specifications require I/O thresholds (Vol, Voh) that differ from those normally adopted on other CMOS I/O. Therefore you will find that a specific pair of pins is normally assigned to the interface (RC0 and RC1).
When the PPS feature is available, the I²C interface *can* be re-routed on any other available digital I/Os but at the expense of full compliance with the bus specifications.
On the PIC12LF1552, for the first time, the MCLR pin has been multiplexed with the I²C SDA function to maximize the use and availability of analog inputs in the tiny 8-pin package.

Homework

- Check out the System Management Bus (SMBus) and its applications in motherboards and Smart Battery Systems.
- Check out the Power Management Bus (PMBus) and its applications in digital management of Power Supplies.
- Compare the I/O (V_{hl} and V_{ll}) specs of an I²C input pin against the typical thresholds that apply to standard CMOS and TLL I/Os.

Online Resources

AN1028 - Recommended Usage of Microchip I²C Serial EEPROM Devices
AN1248 - PIC MCU-Based KeeLoq Receiver System Interfaced Via I²C
AN1302 - An I²C Bootloader for the PIC16F1
AN1365 - Recommended Usage of Microchip Serial RTCC Devices
AN1488 - Bit Banging I²C on Mid-Range MCUs with the XC8 C Compiler
AN1630 - USB to I²C Bridge Reference Guide
AN690 - I²C™ Memory Autodetect
AN734 - Using the MSSP Module for Slave I²C Communication

SPI – Synchronous Port

Introduction

The Serial Peripheral Interface is perhaps the simplest of the serial interfaces as it is essentially composed of just two shift registers (one at each end) swapping data. Its main disadvantage is that, contrary to the I²C, it requires a larger number of pins/wires and this number increases with the number of devices added to the bus.

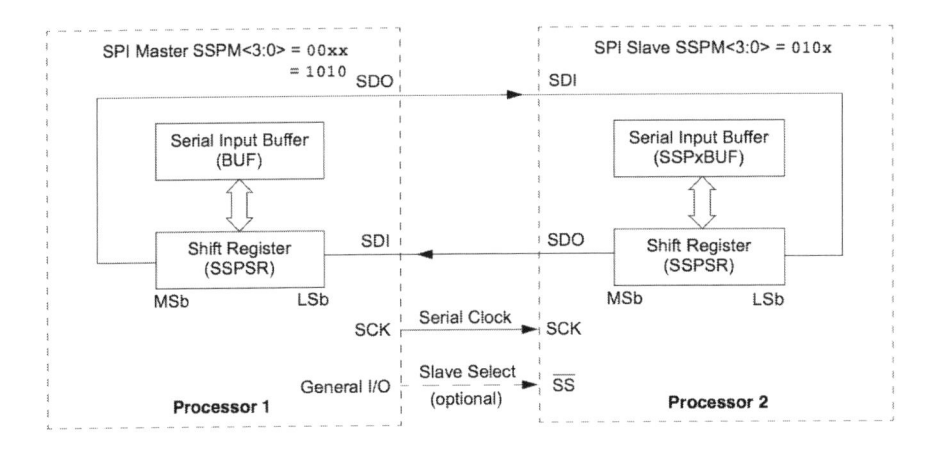

Figure 6.2: SPI Master/Slave connection

How it works

All PIC16F1 microcontrollers that support the SPI interface do so via the Master and Slave Synchronous Port (or MSSP) peripheral. This is shared with the I²C interface (see previous section) to save on configuration registers and buffers in typical Microchip frugal approach.

Master and Slave operation are supported (the distinction being which unit does provide the serial clock SCK) but also all four possible modes accounting for the possible permutations of the active clock edge (rising or falling) and clock polarity.

Note that because of its symmetrical nature, each data transfer on the SPI port is simultaneously a transmission and a reception.

Applications

The use of SPI interfaces is typically restricted to the communication with devices that are located in close proximity, on the same PCB. This includes sensors, displays, non volatile memory devices, radio modules, transceivers, but also communication with other microcontrollers located in the vicinity.

The synchronous nature of the serial port means that it is not necessary to use high accuracy (crystal) oscillators reducing the cost of the application.

The biggest advantage is perhaps the data rate that is often limited only by the clock speed of the PIC microcontroller, when it could otherwise extend up to 20 Mb/s.

A notable class of applications for the SPI interface is composed by the MMC/SD cards, i.e. mass storage (Flash) devices. In fact a SPI mode is always offered on all such devices as a minimum common interface for all sizes and formats.

Limitations

The SPI interface can operate while the device is in sleep, in fact when configured in Slave mode, it will wake up the device as new data is received. But the SPI interface still requires a bit of hand holding to complete an entire data transaction with an external device (serial EEPROM or sensor), so we cannot truly claim core independent operation (yet).

Since several registers are shared between the SPI and I^{2}C function, once the MSSP peripheral is used for one of the two functions, it is not possible to use the other. So if you need both I^{2}C and SPI interfaces in an application, you will be looking for devices with two indepenedent MSSP peripherals.

MCC API

The MCC distinguishes the SPI interface configuration between Master and Slave mode, but contrary to the I^{2}C case, makes no provision for use of interrupts.

Instead, the code generated (placed in the *spi.c* file) contains two main functions: SPI_Exchange8bit() and SPI_Exchange8bitBuffer() that allow transmit/receive of a single byte or a short buffer of data at once. A small corollary of *getters*, that check for overflow and other error conditions, complete the picture.

Example

The example below is a typical case of SPI interface to an external serial EEPROM model 25LC256 (32 kByte).

```c
/* Project:    SPI Master
 * Device:     PIC16F1829
 */
#include "mcc_generated_files/mcc.h"

// 25LC256 Serial EEPROM commands
#define SEE_WRSR   1      // write status register
#define SEE_WRITE  2      // write command
#define SEE_READ   3      // read command
#define SEE_WDI    4      // write disable
#define SEE_STAT   5      // read status register
#define SEE_WEN    6      // write enable

main()
{
    uint8_t   status, data;
    uint16_t  address = 1234;

    SYSTEM_Initialize();

    while( 1)         // main loop
    {
        // 1. send a Write Enable command
        CSEE_SetLow();                    // select the Serial EEPROM
        SPI2_Exchange8bit( SEE_WEN);      // send write enable command
        CSEE_SetHigh();                   // deselect, terminate command

        // 2. Check the Serial EEPROM status
        CSEE_SetLow();                    // select the Serial EEPROM
        SPI2_Exchange8bit( SEE_STAT);     // send a READ STATUS COMMAND
        status = SPI2_Exchange8bit( 0);   // send/receive
        CSEE_SetHigh();                   // deselect, terminate command
        // expect status == 2

        // 3. Read a byte from EEPROM address
        CSEE_SetLow();                         // select the Serial EEPROM
        SPI2_Exchange8bit( SEE_READ);          // send a READ STATUS COMMAND
        SPI2_Exchange8bit( address>>8);        // send address MSB
        SPI2_Exchange8bit( address);           // send address LSB
        data = SPI2_Exchange8bit( 0);          // get data
        //... (or more)
        CSEE_SetHigh();                        // deselect, terminate command

    } // main loop

} // main
```

PinOut

Early PIC16F1 devices will have a fixed assignment for the MSSP peripheral where SPI and I²C I/Os are typically overlapping (Port C, pin RC0 through RC3). More recent devices featuring the PPS are able to re-route the SPI interface I/Os to any available digital pin.

Homework

- Compare the SPI module with the EUSART Module when used in Synchronous mode.
- Look in the Microchip Library for Applications, in the File System library (MDD File System in the legacy edition) for examples of use of the SPI port to access data on an MMC/SD card.
- Notice how all signals on a SPI bus are uni-directional. Consider how this can simplify design in systems that require Galvanic isolation.
- Look up the definition of the Microwire interface and consider if/how this could supported perhaps with use of the PPS module.

Online Resources

AN1029 - Recommended Usage of Microwire Serial EEPROM Devices
AN1040 - Recommended Usage of SPI Serial EEPROM Devices
AN1245 - Recommended Usage of SPI Serial SRAM Devices
AN767 - Interfacing Microchip's Fan Speed Controllers to a SPI

EUSART – Asynchronous Serial Port

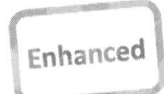

Enhanced

Introduction

Universal Asynchronous Receivers and Transmitters (UARTs) used to rule the embedded control world. They represented the one and often only way to connect to a PC, or a terminal but also a printer and, through a modem, the world! Today, terminals and modems belong positively to the archeological museum of computing history. PCs have long stopped offering the once ubiquitous 9-pin D-type connector and replaced it with ever faster USB ports. But the asynchronous ports continue to find new applications thanks to their simplicity and flexibility. In fact PIC16F1 devices feature an *Enhanced* UART that can perform new tricks including: automatic baud-rate detection (used by the *LIN-bus* standard) and *Synchronous* communication, hence the new acronym (EUSART).

How it works

The asynchronous nature of the serial port implies the absence of a clock signal, otherwise present in the I²C and SPI protocols we have seen in the previous sections. This comes with the benefit of a reduced number of pins/wires but is otherwise balanced by an increased oscillator cost as a precise bit timing (baud rate) must be agreed a priori between the two ends being connected.

Once the baudrate is established, (either fixed or by applying the auto baud rate detection mechanism) the falling edge of the start bit is all that is required to synchronize a receiver and a transmitter for the following eight bit time slots. A stop bit (eventually double) provides the closing of each individual characters transmission and a first primitive check of data integrity. A ninth data bit was originally included in the protocol as a second means of data integrity check by providing the *parity*, ensuring that each character transmission would eventually contain an even number of 1s and 0s.

But in modern applications the ninth bit is often used for other, better purposes. For example in the *RS485* protocol, which transforms what is otherwise a point to point connection into a proper bus, the ninth bit is used to distinguish the first byte (containing the destination address of a new message sent by the master) from the following pieces of data.

Applications

Traditional RS232 standard communication (requiring the adoption of a particular transceiver capable of translating the signaling to ±12V) are still used occasionally to provide module to module communication over long-er distances than otherwise possible with SPI and I²C.

RS232 standard communication can still be used occasionally to connect an embedded application with a Personal Computer by using a Serial to USB dongle, occasionally implemented using a device capable of bridging between the two serial interfaces (such as those in the PIC16F145x family).

RS422 standard communication (requiring a more complex fully balanced signaling transceiver) allow even longer distances (up to hundreds of meters) but are becoming increasingly rare.

RS485 standard bus connections have a huge legacy in industrial control applications where they are used to create long multi-drop chains of devices. LIN-bus standard communications require yet another specialized transceiver and are preferred in automotive applications in non-critical systems where reduced cost is favored over connection speed and data integrity.

Lighting applications can use UARTS to implement the DMX512 protocol.

Limitations

Since the only opportunity for receiver and transmitter is to synchronize on the start bit and the timing cannot differ by more than half a bit after the first stop bit, it can be proven that the maximum oscillator tolerance allowed on each device is ±2.5% when using 10 bits (start bit + 8-bit data + stop bit) and ±2.27% when using 11 bits (start bit + 9-bit data + stop bit). The error can be allowed to double to ±5% and respectively ±4.55% when one of the two devices has a known *accurate* oscillator and the total system clock error is contained.

This level of precision can be achieved by most internal oscillators available on PIC16F1 devices thanks to factory calibration at room temperature, but is otherwise hard to maintain over the typical large operating temperature range (-40 to +125°C) and even more so if the supply voltage is not regulated accurately. When wide operating temperature ranges and/or voltages are expected, it is wiser/safer to use an external resonator/crystal oscillator or to make use of the *auto-baudrating* capabilities of the EUSARTs.

The maximum data transfer speed of an asynchronous serial interface is limited in most cases by the distance and the quality of cabling used. Even when communicating over short distances, speeds in excess of 115k baud are rare, with DMX512 being perhaps the most notable exception (250k baud).

MCC API

The MCC provides support for the EUSART module by providing the basic configuration function `EUSARTx_Initialize()` and a pair of straightforward `EUSART_Read()` and `EUSART_Write()` functions. If the STDIO redirection option is checked, these will be wrapped in a `getch()` and `putch()` pair to allow the XC8 compiler to use them and to implement the full `printf()` capabilities.

Further, if the *interrupt support* option is checked, the interrupt module is added automatically to the project and both read and write functions are fully buffered.

Example

The following example demonstrates the use of the EUSART with STDIO redirection to communicate with a serial console:

```
/* Project: Console
 * Device:  PIC16F1709
 */
#include "mcc_generated_files/mcc.h"

main()
{
    char c;

    SYSTEM_Initialize();

    while ( 1)
    {
        // 1. text prompt
        printf( "Is this Rocket Science?:(y/n)");

        // 2. read a character
        c = getch();
        puts("");
        if ( c == 'y')
            puts( "No, this is too easy!");
        else
            puts( "That's what I am saying!");

    } // main loop
}// main
```

PinOut

The EUSART was originally offered on devices without PPS where only a limited selection of pin was available (TX-RC5, RX-RC4). More recent microcontroller models added the PPS flexibility to expand the number of options to any available digital I/O.

Homework

- Investigate how the Synchronous mode works (the S in EUSART).

- Consider using the CRC module in combination with the EUSART to accelerate support of the LIN bus protocol.

- How would you implement Manchester encoding/decoding of the data stream to/from the EUSART? (Hint: consider using the CLC)

- How would you apply 38KHz typical IR modulation to the data stream output of the EUSART? (Hint: look at the CLC or DSM modules)

Online Resources

TB3069 - Use of Auto-Baud for Reception of LIN Serial Communications

AN1659 - DMX512A
AN1099 - LIN 2.0 Compliant Driver Using the PIC16 Microcontrollers
AN1310 - High-Speed Bootloader for PIC16 and PIC18 Devices
AN1465 - Digitally Addressable Lighting Interface (DALI) Communication

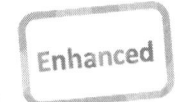

USB - Universal Serial Bus - Active Clock Tuning

Enhanced

Introduction

The Universal Serial Bus has supplanted a number of older interfaces common to the original PC peripherals world including floppy disk interfaces, parallel printer ports and, most importantly, serial ports.

PIC microcontrollers had initially been supporting the USB bus with Low Speed interfaces (up to 1.2 Mbit/s) aiming at covering the roles of simple keyboard controllers and other Human Interface Devices (HID), but with the PIC16F1 generation the USB module capability has been expanded to include High Speed 2.0 applications (up to 12 Mbit/s). Perhaps even more important has been the addition of the Active Clock Tuning technology that makes it possible for the PIC16F145x family devices to connect in High Speed mode using exclusively an internal oscillator!

How it works

The USB serial interface engine (SIE) is a relatively complex peripheral module that performs complete transactions of relatively large blocks of data (up to 64 bytes of data at a time when operating in *bulk* mode) from PIC to Host and vice-versa. It would be tempting to claim it to be a Core Independent Peripheral but the complexity of the USB protocol is such that there is still a significant amount of handling and oversight required by the MCU core.

Luckily the Microchip Library for Applications (MLA), contains a very complete USB library with ready to use support modules and examples for all the most common classes of USB applications including:
- HID, the human interface device class composed of keyboards, mice, digitizers ...
- CDC, the communication device class, covering the all important *serial port emulation*

- Audio, covering microphones and loudspeakers
- Printer, covering the connection to printing devices
- Mass Storage, covering USB memory sticks and hard drives

Active Clock Tuning

The most interesting feature of the PIC16F145x family is the ability to connect up to Full Speed without using any external crystal oscillator and therefore providing a very fast and economical link. This would seem impossible if we consider the Full Speed bit-rates (12Mbit/s) and the tolerances required by the standard for such connections (better than 0.25%).
PIC16F145x internal oscillators are not that accurate, but they are *tunable*. Their frequency can be adjusted with very fine granularity via a control register known as the OSCTUNE.
The Active Clock Tuning technology makes use of this feature and another peculiarity of the USB protocol. In fact, you might know that the USB protocol wants the host (the PC) to start every communication sequence (a *frame* in USB lingo) with a particular *start of frame* (SOF) *token*. Frames must be produced at "exact" intervals of one millisecond and this timing, being provided by the host, is supposed to be extremely accurate.
The Active Clock Tuning technology uses the SOF token detection mechanism to derive a reference clock signal and to act on the internal oscillator (tuning register) to synchronize it. As temperature and supply voltage change, the ACT module re-calibrates the internal oscillator maintaing the required tolerances and a stable USB communication.

Applications

The list of USB connected applications has expanded well beyond the original list of PC peripherals the USB bus was designed to supplant. By supporting both Low Speed and Full Speed connections, PIC16F1 devices can cover a very large number of new applications with a particular preference for those that require low power consumption, small size and very low cost (thanks to ACT).
The simplicity of implementation of a customizable UART to USB bridge (serial port emulation) and the flexibility of the HID class for quick and dirty data and command transfer to/from a PC, make these two the main PIC16F145x family use cases.

Limitations

The USB protocol can be intimidating even though the MLA library provides much of the functionality *out of the box*. Although the protocol is theoretically capable of 12Mbit/s, only a fraction of that bandwidth is available to each single device on the bus at any time. Further considering the protocol overhead, it is not realistic to expect effective sustained transfer rates in excess of 1 Mbit/s. RAM memory size can be another limiting factor in certain classes of applications such as mass storage and/or printer device.

MCC API

As of this writing, the MCC is not (yet) capable of assisting with the project configuration. This is supposed to change with the future editions of the MCC tool as complex communication frameworks will be added.

The Microchip Library of Applications (MLA), contain all the support firmware required for the most popular classes of USB applications (HID, CDC, Digitizer, Mass Storage...) and provides source code that is compatible with all PIC architectures (PIC16, PIC18, PIC24, dsPIC, PIC32).

PinOut

Similarly to other serial communication interfaces, USB applications require use of special transceivers that are integrated on chip but only on fixed pin locations. Also pull ups for the selection of the connection speed are provided internally by the PIC16F145x devices.

Homework

- Look at the Microchip Library for Applications demo projects featuring the PIC16F145x family (targeting the Low Pin Count USB board).
- USB developers need a licensed (USB_IF) unique Vendor ID (VID) number, to ensure that their applications are matched with the right drivers. If your application is not going to exceed the 10,000 units in volume, in order to avoid steep licensing fees, you can ask Microchip to use a shared VID (among all customers) and to be assigned a unique Product ID (PID)

Online Resources

https://www.microchip.com/usblicensing – USB VID/PID licensing

AN1546 - USB Keypad Reference Design

AN956 - Migrating Applications to USB from RS-232 UART with Minimal Impact on PC Software

AN1630 - USB to I²C Bridge Reference Guide

https://code.google.com/p/pic16f1454-bootloader – An Open Source USB Bootloader example project hosted on Google Code.

Example

The following example demonstrates the use of the HID Custom class (MLA USB library) to implement a simple "sensor":

```
/* Project: USB HID Sensor
 * Device:  PIC16F1459
 */

#include "system.h"
#include "USB/usb.h"
#include "USB/usb_function_hid.h"

void DecodeCMD( void)
{
    switch( RxBuffer[0])        // decode first byte
    {
      case 0x37:            // interpret as "Read POT" command
        if ( !HIDTxHandleBusy(USBInH))   // ready to send packet back
        {
            WORD_VAL w;
            w.Val = ADC_GetConversion( Potentiometer);
            TxBuffer[0] = 0x37;    // Echo back
            TxBuffer[1] = w.v[0];  // LSB
            TxBuffer[2] = w.v[1];  // MSB

            // return transmit buffer to USB SIE
            USBInH = HIDTxPacket(HID_EP, &TxBuffer[0], 64);
        }
        break;

        // add more commands here...
      default:
        break;
    } // switch
}
```

```
int main(void)
{
    SYSTEM_Initialize();

    while(1)
    {
        USBDeviceTasks();

        if ( USBDeviceState == CONFIGURED_STATE)
        {
            if ( !HIDRxHandleBusy(USBOutH))      // packet received
            {
                DecodeCMD();

                // return receive buffer to USB SIE
                USBOutH = HIDRxPacket(HID_EP, (BYTE*)&RxBuffer, 64);
            } // if received
        } // if configured
    } // main loop
} // main
```

Chapter 7 - Analog Functions

Comparators

Enhanced

Introduction

An analog comparator can be quite a simple (economical) yet helpful module. While it certainly cannot replace the convenience of an Analog to Digital Converter, in many applications, when a simple threshold needs to be detected, it can provide a fast and core independent response in a fraction of the time. The accent is once more on the *core independence*, that is the ability to drive with its output the input of other peripherals, generating a chain of events that is fast and does not depend on the CPU presence and current workload.

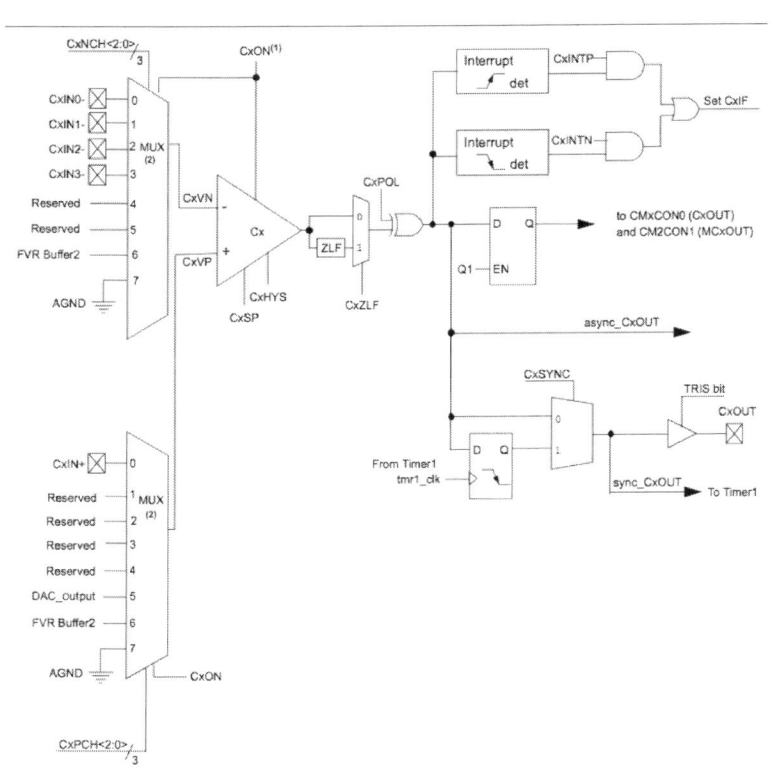

Figure 7.1: Comparator Block Diagram

How it works

The typical comparator modules found on the PIC16F1 family devices are equipped with two input multiplexers (one for each input side). The FVR buffered output can be present as an option on one side (or both depending on models) and the digital output can be published on a device pin if desired or can be routed internally to other modules.
Comparators are also commonly paired with DAC modules, the most common being a basic 5-bit DAC, but in more recent models 8-bit DACs can be found.
An interrupt event can also be generated (on rising, falling or both edges of the output signal) and the same event can be used as a wakeup trigger for the microcontroller.

Applications

A comparator is a fundamental building block of many mixed signal circuits, including:

- Over-voltage and under-voltage thresholds detection in power supply applications.
- Over-current detection in motor control applications.
- Low Battery detection.
- Many Switch Mode Power Supply topologies make use of a comparator (to detect peak current for example).
- Many Analog to Digital Conversion circuits are built using comparators, but most practical applications will rather use the available internal ADC module instead.

Limitations

Standard comparator modules offer a compromise between power consumption, cost and speed. As in many similar cases the result is that the generic comparators offered on general purpose microcontrollers cannot excel at any of the three. In particular the speed is limited to approx 1us, while the current consumption is typically 50uA. Even a low power mode optionally available will reduce that current consumption to a typical value of 10uA. Clearly not an XLP function. You might want to remember to disable them before entering sleep mode.

High Speed Comparators

Enhanced

In more recent devices designed for power supply (SMPS) applications, such as the PIC16F17xx families, the comparator response speed has been significantly increased to reach 50ns (a two orders of magnitude improvement). This enables power applications to reach comfortably the 500kHz switching frequency.

MCC API

The MCC offers typically a default `CMPx_Initialize()` function and a simple macro `CMPx_GetOutputStatus()` to test the output of the comparator in software.

PinOut

As for all analog functions, pin mapping of comparator inputs is considerably more limited than that of digital functions. The negative input of the comparator offers a few more choices of external pins, while the positive input is constrained to a single pin option. The (digital) output of the comparator is also optionally available to the outside world via the PPS (when available).

Homework

- Check the reaction speed of a comparator and contrast it with the time required for an ADC conversion, add MCU interrupt response time.
- Compare reaction time of normal comparators vs. high speed comparator specifications in PIC16F17xx devices.

Online Resources

AN1427 - High-Efficiency Solutions for Portable LED Lighting
AN1463 - NiMH Trickle Charger with Status Indication
AN1384 - Ni-MH Battery Charger Application Library
AN1138 - A Digital Constant Current Power LED Driver
AN874 - Buck Configuration High-Power LED Driver
AN700 - Make a Delta-Sigma Converter Using a Microcontroller's Analog Comparator Module

DAC – Digital to Analog Converter

Enhanced

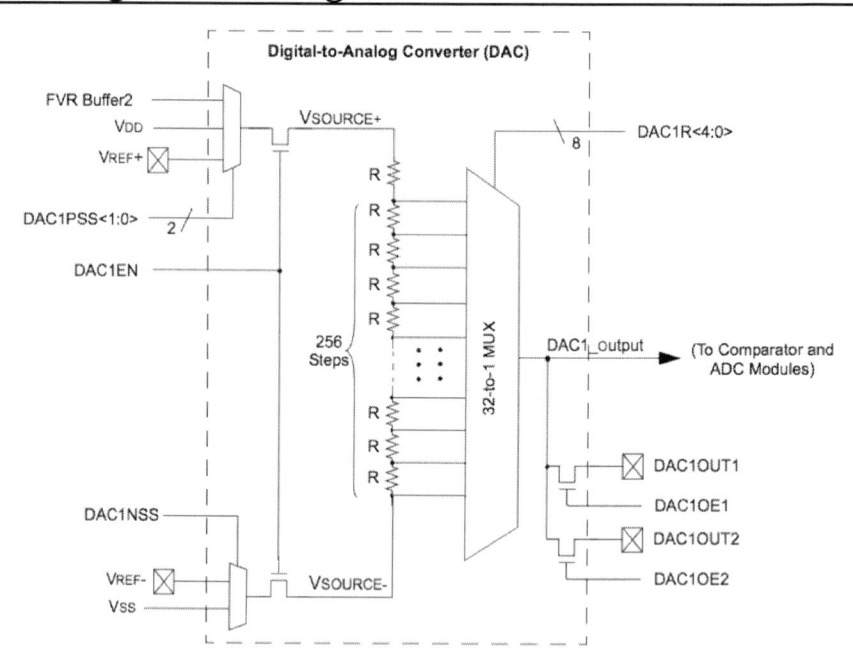

Figure 7.2: DAC Block Diagram

Introduction

For the longest time PIC16 microcontrollers have been featuring a very simple DAC module and only as part of the Analog Comparator block. In all those cases the resolution was limited to 5 bit which provided just enough flexibility to implement simple programmable threshold detection circuits. With the introduction of the PIC16F170x/1x the DAC has been extended to 8 bit of resolution and with the PIC16F176x family it has been increased further to 10 bit.

How it works

All DAC models, regardless of the resolution (N), share the same basic design: a simple resistor ladder with 2^N elements of approximately 600 Ohm. An analog multiplexer picks the desired tap in the ladder.

Two more analog multiplexers are offered at the top and (occasionally) the bottom of the resistive ladder. The number of input options vary with the PIC model but always include the possibility to connect to selected external pins (labeled Vref+ and Vref-).

By default the top of the ladder is connected internally to the device Vdd (and the bottom to Vss) which covers the simplest use case: the output of an analog value as a fraction of the device power supply.

The DAC output circuit typically includes the option to present the analog voltage output on one or two pins but provide also connections to multiple other analog modules inside the microcontroller such as the ADC, via a dedicated channel and any number of comparators. Internal connections can help save pins, but are also very important to keep noise in check.

Enabled DAC output pins get their digital I/O functionality disabled. If you try to *read* the pin input value (PORTxbits.Rxy) you will always get 0.

Applications

When the multiplexer at the top of the DAC ladder offers the option to connect to the Fixed Voltage Reference (FVR), the DAC can be used to generate an output voltage that is *absolute*. This can be obviously useful in battery applications to provide accurate thresholds, independent of the battery charge.

In power supply applications, the top of the DAC ladder can be connected to the external (Vref+) pin and if this is in its turn connected to the AC input voltage (110-220V) via an adequate partition, the output of the DAC can be used as a threshold that tracks the input sinusoid and allows the power application to simulate a resistive behavior or in other words to provide Power Factor control.

> **TRICK**
>
> When both the top and the bottom of the ladder can be routed to external pins (on some models only the top), it follows that the DAC can be used as a very flexible programmable load for any analog input signal. In fact when both the top and bottom are routed to the Vref+ and Vref- device pins, it is possible to use the DAC as a floating programmable resistor!

Limitations

Obvious limitations apply to the use of the DAC inputs and outputs when you consider the constraints of a typical microcontroller design:

- Never allow Vref+ to exceed Vdd
- Never allow Vref- to fall below Vss
- Consider the impedance of the resistive ladder circuit (which is always pretty high). Buffer the output before applying pretty much ANY load to it. You could use one of the integrated operational amplifier modules (configured for unity gain) available on selected PIC16F1 models.

MCC API

The MCC makes configuring the DAC a trivial matter generating a *dac.c* source file containing a default `DAC_Initialize()` function.

In case the FVR is selected as the Positive Reference (top of the ladder), the MCC will add automatically the FVR resource to the current project and will include its initialization in the default `SYSTEM_Initialize()` function.

In case any of the output pins (`DACOUT1` or `DACOUT2`) are selected, the MCC will require you to lock the corresponding pin in the Pin Manager window.

Example

The example below generates a simple test triangular waveform.

Analog Functions - 113

```
/* Project: DAC Triangular Waveform
 * Device:  PIC16F1708
 */
#include "mcc_generated_files.h"

main()
{
    uint8_t count=0;

    SYSTEM_Initialize();

    while(1)
    {
        for(count=0; count<=255; count++)
            DAC_SetOutput( count);
    }
}
```

Pinout

The DAC output is offered multiplexed on two fixed pins. Notice that one of the two options is shared with the in-circuit programming interface (ICSP) CPD pin. Special considerations might be required to ensure that this pin is not *loaded* in a way that would interfere with proper ICSP operation.

Notice also how when the top and bottom of the DAC ladder are connected to external pins (Vref+ and Vref-) they compete for the same positions. When all three points of the ladder are required to be accessible from the outside of the device, the only option is then to accept the overlap with the ICSP interface.

Homework

- Compare the design of the standard DAC module with 5 or 8 bit of resolution with that of the PIC12F752. Notice the *focused range* mode designed to best address power supply applications by providing the equivalent resolution of a 9-bit DAC without the cost.

FVR – Fixed Voltage Reference

Introduction

While most analog circuits on board of a microcontroller can and use the supply voltage (Vdd) as a reference, having an independent a fixed voltage reference module on board can be of great convenience in a number of analog/mixed signal applications. The FVR module as implemented on the PIC16F1 family of microcontrollers attempts to strike a difficult balance among three conflicting objectives: stability, low power consumption and low cost.

How it works

The FVR is composed of a basic voltage reference circuit (a band-gap) that produces an output of 1V with a relatively high impedance. This is followed by one or more buffers that decouple the source from the circuits using it (DAC, Comparators, ADC input, ADC reference) optionally introducing a gain (1x, 2x, 4x) to produce one of three possible output voltages 1 V, 2 V, 4V.

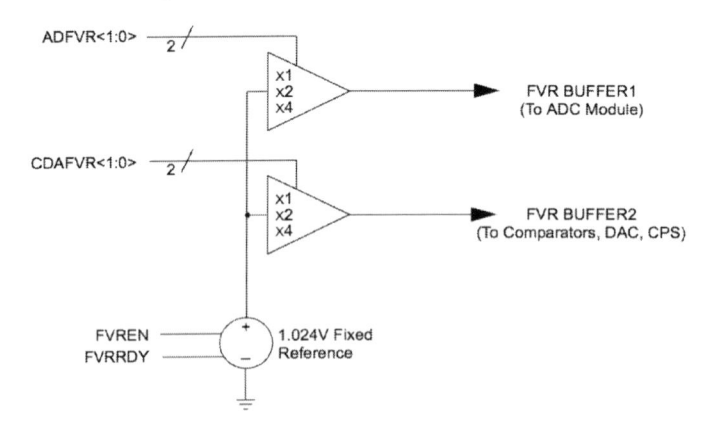

Figure 7.3: Fixed Voltage Reference Block Diagram

Applications

Having a fixed voltage reference is important in all battery applications and/or whenever the supply voltage is not regulated, and in general not known. Even when a voltage regulator is used (LDO), its typical accuracy (+/- 10% is a common value) might be insufficient for the application whereas a more accurate regulator might imply a higher cost.

In general a fixed voltage reference will provide greater immunity from the noise present on the power supply rails of the application and therefore a better (cleaner) analog signal chain.

Limitations

As mentioned in the introduction, the FVR offered in the current generation PIC16F1 devices was designed to compromise between cost, stability and low power consumption. As a result is cannot fit the tight boundaries of an eXtreme Low Power specifications. Operating currents between 10uA and 70uA are common, making it mandatory to ensure the peripheral is turned off before entering a low power mode. Consider also other circuit that require the FVR (implicitly) for their internal working, including: HFINTOSC, BOR, and the internal LDO.

Further, to achieve the target accuracy the FVR circuit (amplifiers) require a stabilization period that can last in excess of 30us.

Lastly, the typical accuracy of the FVR (over temperature and supply voltage) is limited to one or two points of percentage. This must be accounted for when using the 10-bit ADC, as the impact on the measurement will be considerably higher than the ADC own resolution limitations (offset and gain errors). Although this might sound disappointing, to be fair if we are to compare the FVR with a precision voltage reference (external discrete device), which would provide higher accuracy and stability, we must recognize that it would come at a price higher than that of the microcontroller itself.

MCC API

The MCC adds automatically the FVR to a project whenever its output is used as a reference for the ADC or DAC, or as an input for one of the ADC channels or for a comparator. The API generated includes only two functions: the default FVR_Initialize(), and FVR_IsOutputReady() to ensure the FVR is up and running and the stabilization period is past.

Example

This short example sequence might be used upon device wakeup from a low power mode where it might have been disabled:

```
...
FVR_Initialize();                  // enable the FVR
while( !FVR_IsOutputReady();       // ensure FVR output stable
sensor = ADC_GetConversion( InputChannel);
...
```

PinOut

The FVR output is meant to be connected *internally*, but could be presented on an external pin via the DAC module. Use the FVR as the top reference and set the DAC output to its maximum value.

Homework

- What would happen if you selected the 4x output buffer multiplier when Vdd = 3V?

- Investigate FVR accuracy and available options for external voltage references.

- Compare FVR accuracy with typical LDO output specifications.

ADC – Analog to Digital Converter

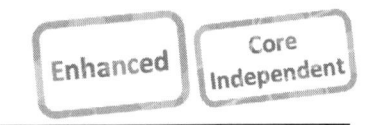

Enhanced *Core Independent*

Introduction

A solid Analog to Digital Converter module with 10-bit of resolution is pretty much standard equipment on all PIC16F1 microcontrollers. As of this writing, only selected models (PIC16F178x family) offer 12-bit resolution ADCs.

There are two important features of this module that make it stand out among competing devices: its excellent *accuracy* and the large number of input *channels*.

Analog Functions - 117

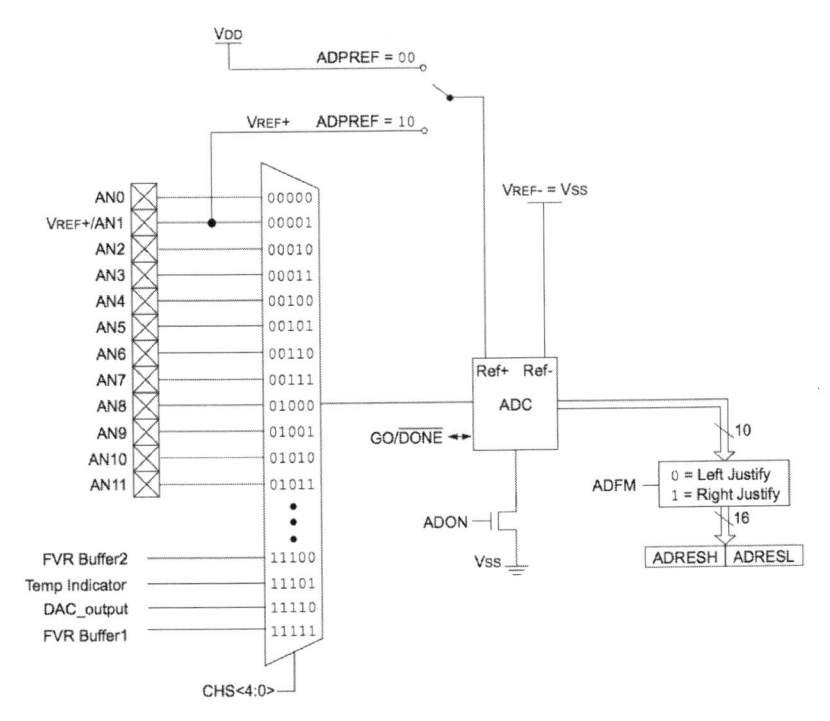

Figure 7.4: Analog to Digital Converter Block Diagram

In fact *accuracy* and *resolution* are often confused, where in reality a high(er) resolution ADC does not necessarily provide (more) accurate readings. What is more accurate: a 10-bit ADC with ±1 LSB error or a 12-bit ADC with ±6 LSB error? (Hint: compare the ENOB!)

The large number of input channels is one of those features that help the design flexibility. It can be used to accommodate more sensor inputs but also to simplify routing and help keep separation between analog signals and noisy digital pins thereby improving the application performance.

How it works

The traditional PIC16F1 ADC module is of the successive approximation register type (SAR), which allows for a conversion speed (~100K smps) that is adequate for the processing capabilities of an 8-bit microcontroller.

The sample and hold circuit is of the simplest possible design, basically consisting of a small (internal) capacitor connected directly to the input (via the mux). When the GO bit (in the ADCON register) is set or an *auto-conversion* trigger input is received, the S&H capacitor is disconnected from the input and the conversion process begins. Upon completion of the conversion, (13 ADC-clock cycles later) the result is available in the ADRES register (16-bit wide) and a flag is set optionally generating an interrupt.

The ADC clock can be derived from the system clock or can be generated internally by an ADC-exclusive low-power internal RC oscillator of fixed frequency (~500kHz). This option allows for ADC operations in sleep, when all other clock sources are stopped and therefore will provide the best accuracy. If the application does not allow for the microcontroller to sleep during conversion though, it is best to use the system clock derived option. In this way while the noise will be higher, it will be at least *correlated* with the sequencing of the ADC internal operations and therefore will produce more repeatable and stable results.

Core Independent

Auto-Conversion Trigger

The auto-conversion trigger (selected by the TRIGSEL field in the ADCON2 register) is the key to achieve the synchronization between the conversion and other peripherals. It is thanks to this feature that it is possible to build complete complex chains of events connecting multiple digital and analog peripherals to manage real world events while leaving to the core a pure supervisory role, with minimal power consumption and yet great responsiveness. This is Core Independent Zen!

Limitations

The ADC module has a relatively low input impedance (<1k Ohm) which requires external signal sources to match it (< 5K÷10K Ohm) or to be buffered to avoid significant sampling error. Integrated operational amplifier modules such as those available on the many PIC16F17xx family models (see the OPA module later in this chapter) can be used for this purpose.

In order to increase the measurement resolution when the input signal is available in a limited sub-range of the supply voltage, some applications can benefit from providing external Vref+ and Vref- inputs. Note that this option is effective only as long as the difference between the two references is ≥ 2V. Below such threshold any gain in resolution is negated by a corresponding increase in the conversion error.

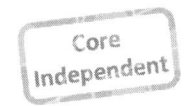

HCVD – Capacitive Sensing

Core Independent

HCVD is a feature of the ADC module found on the most recently introduced models of the PIC16F1 family. The acronym stands for Hardware Capacitive Voltage Divider a reference to the Core Independent implementation (hence the hardware) of one of the fundamental technologies used for capacitive sensing in all its many applications including cap-touch sensing, metal-over-cap, pressure sensing, level sensing and infinite variations of the above.

The basic CVD technique, pioneered more than 20 years ago by Dieter Peter, is based on the most basic principles of physics applied to the parallel of two capacitors but also makes brilliantly use of the peculiar capabilities of PIC I/Os (high current drive, direction control) and the ADC module S&H circuit. Because of these most basic ingredients, the CVD technique has found immediate and successful application on all PIC16F1 microcontrollers enabling a large number of applications in the most disparate fields from automotive to home appliances, lighting and even energy metering.

One of the limitations of the original CVD approach though had been the strict timing requirements that often forced the designer to use optimized assembly language routines and loaded heavily the microcontroller core. The new ADC modules, featuring HCVD technology, enable the entire CVD acquisition to be performed automatically relieving the CPU from the real time constrains and allowing the algorithms to focus on filtering and higher level task while the entire application can now be written in C language.

MCC API

The MPLAB Code Configurator groups conveniently all the ADC module settings in a single dialog window. From there it is possible to select:

- Clock source, checking automatically that proper ADC clock limits are respected
- References to be used, if the FVR is selected, it is automatically added to the configuration
- Input channels, configuring and naming the respective pins in the pin_manager

- Interrupt generation, automatically adding the interrupts_manager module to the application.

The resulting set of functions, placed in the *adc.c* file, include the default initialization, and simple functions to start an acquisition `ADC_StartConversion()`, including channel selection and proper (configurable) acquisition timing, check for termination `ADC_IsConversionDone()` and fetching of the result `ADC_GetConversionResult()`.

Alternatively the entire sequence can be obtained by a single (blocking) call to `ADC_GetConversion()`

MLA – Microchip Library for Applications – Touch Library

The Microchip Library for Applications offers, among other libraries, a rather complete *Touch Sensing Framework* that is compatible with the PIC16F1 architecture and the XC8 compiler.

Example

The following example illustrates, in ten lines of code or so, how it is possible to use the mTouch Framework (part of the MLA) to read eight capacitive buttons and activate one of the corresponding eight LEDs connected to the device PortD. Debouncing, toggle, single/multiple selection parameters can be simply configured via the *mTouchConfig.h* file (not shown).

```
/*
 * Project:        mTouch Demo
 * Device:         PIC16F1937
 * Board:          mTouch eval kit, 8-buttons daughter
 * Requires:       MLA (version 12-15-2013) mTouch Framework
 */
#include "mTouch.h"

void ADC_ISR(void)
{
      mTouch_Scan();              // mTouch timer interrupt
}
```

Analog Functions - 121

```
void main(void)
{
    int8_t  i;

    SYSTEM_Initialize();              // init the I/Os
    mTouch_Init();                    // mTouch Initialization
    INTERRUPT_GlobalInterruptEnable();

    while(1)
    {
        if (mTouch_isDataReady())    // Button Pressed
        {
            mTouch_Service();        // Decoding
            for( i=0, LATD=0xFF; i<8; i++)
                if ( mTouch_GetButtonState( i) >=  MTOUCH_PRESSED)
                    LATD ^= (1<<i); // PORTD pins are connected to LEDs
        }
    }
} // main
```

PinOut

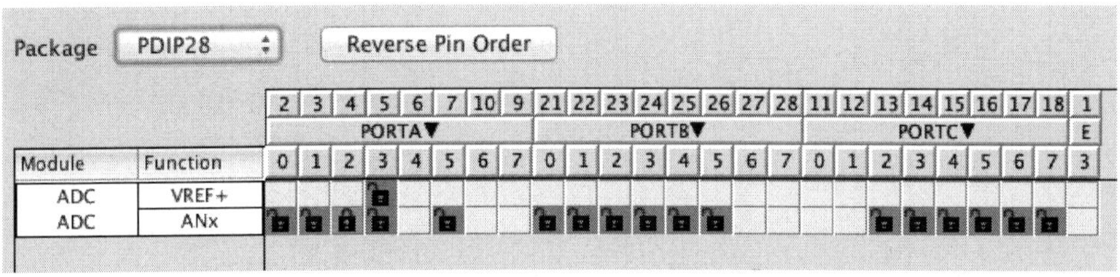

Figure 7.5: Pin Manager, Analog Input Selection

All PIC16F1 microcontrollers offer a very large selection of pins that can be configured as analog input channels to the ADC module.
The Vref+ (and Vref- when available) are fixed in position. See Figure 7.5 for an example based on the PIC16F1513.

Homework

- Compare the definition of Accuracy and Resolution of an ADC converter. Does more resolution imply more accuracy? Why not?

- Consider the possibilities offered by the ADC ability to run while the microcontroller is in sleep, awakening the core only when a complete measurement has been performed.

Online Resources

AN1478 - mTouch Sensing Solution Acquisition Methods: CVD
AN1492 - Microchip Capacitive Proximity Design Guide
AN1334 - Techniques for Robust Touch Sensing Design
AN1325 - mTouch Metal Over Cap Technology
AN1334 - Techniques for Robust Touch Sensing Design
AN1152 - Achieving Higher ADC Resolution Using Oversampling
AN1560 - Glucose Meter Reference Design
AN1626 - Implementing Metal Over Capacitive Touch Sensors
AN693 - Understanding A/D Converter Performance Specifications
AN699 - Anti-Aliasing, Analog Filters for Data Acquisition Systems

Temperature Indicator *New*

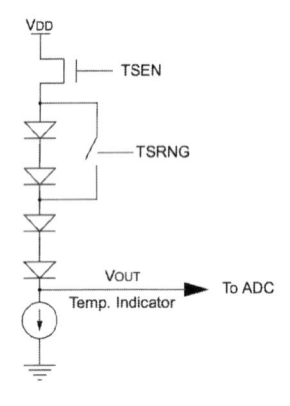

Figure 7.6: Temperature Indicator Block Diagram

Introduction

The temperature indicator is another simple analog module that is characteristic of all PIC16F1 devices. Out of the box, it can provide an approximate temperature threshold detection, but accurate measurements do require some form of calibration. It is for this reason I believe that, with some modesty, the module has been called an *indicator* rather than a proper *sensor*.

How it works

As illustrated in Figure 7.6, a constant current source is connected to a series of four junctions. The output is directly offered as an input to one of the ADC channels.

Note that two of the junctions in the series can be shorted for applications that operated below 3.6V, but also as an option to eliminate the dependency from Vdd from the resulting measurement.

Applications

The value measured is an indication of the temperature of the environment of the microcontroller and if properly placed could be assumed to be close to that of other parts of the embedded application being monitored (motor, power device, ...).

Limitations

Because of the direct dependency of the measurement from Vdd, the Temperature Indicator can achieve only a relatively low accuracy (+/-5°C) unless a two point calibration procedure is followed at the end of the final product assembly and test line.

MCC API

Since the Temperature Indicator shares its control register with the Fixed Voltage Reference, the MCC includes an option to enable the Temperature Indicator inside the same dialog window where the FVR is configured (see Figure 7.7). The selection is therefore reflected in the generated default FVR_Initialize() function.

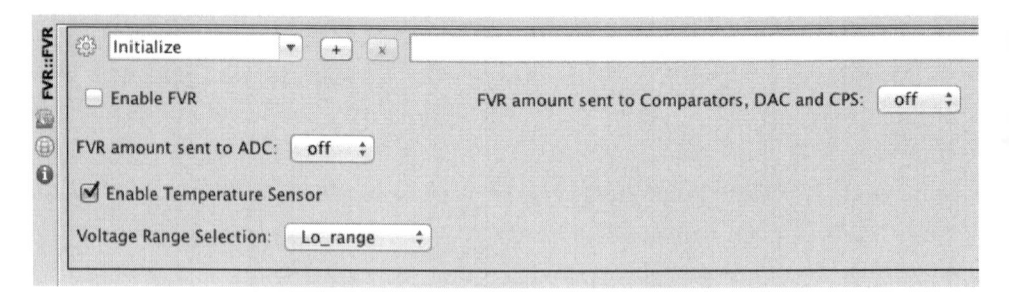

Figure 7.7: Temperature Indicator

PinOut

The Temperature Indicator is directly connected internally to the ADC module. It has no access to any of the device pins.

Homework

- Compare the convenience of an on chip temperature sensor vs. an external (analog or digital) sensor, but also the implications of the thermal resistance of the device package and distance from target.
- Compare the achievable uncalibrated accuracy (typically ±10°) vs. the expected accuracy of external devices such as the MCP9800 and MCP9700 families.

Online Resources

AN1333 - Use and Calibration of the Internal Temperature Indicator
AN1001 - IC Temperature Sensor Accuracy Compensation

OPA – Operational Amplifiers

Introduction

The addition of operational amplifiers, a feature of the PIC16F17xx families, expands significantly the mixed signal capabilities of those devices.
Operational amplifiers make possible to integrate a larger portion of the analog signal chain, reducing cost and in many cases increasing the system performance and reliability.

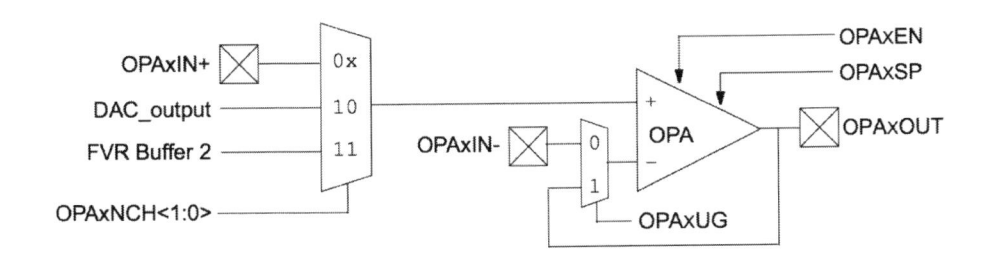

Figure 7.8: OPA Block Diagram

How it works

As seen in Figure 7.8, the OPA modules exposes all three pins of a discrete operation amplifier but also allows some direct internal connection options.
The positive (non inverting) input is equipped with a small multiplexer that can gives access to the fixed voltage reference (buffered) or the DAC output. The inverting input can be connected internally only with the operational amplifier own output creating a unitary gain configuration that can be used to buffer input signals and save a precious pin.

Applications

Thanks to the integrated operation amplifiers it is now possible to condition typical sensor inputs to the ADC by performing:

- *Anti-aliasing* filtering.

- Amplifying and adding or subtracting offset to increase the resolution of measurements.

- Buffering input sources that would otherwise have a too high impedance, therefore speeding up the measurement.

- Provide a differential input measurement capabilities

Limitations

Integrated operational amplifiers accept rail to rail input and are modeled after the MCP60x series of general purpose discrete operational amplifiers featured in Microchip Analog products catalog. The gain bandwidth product is in excess of 2MHz (typical) with a CMRR of 70dB (typical). Unfortunately, a consequence of the integration with the microcontroller structures (I/O drivers) and the added multiplexers, the input offset (although calibrated in production) can reach 9mV (max). The operational amplifiers are also definitely not ideal for XLP applications. A typical current consumption of 350uA (at 5V) is eventually pushed to 600 uA plus when the maximum is considered at high operating temperature. A low gain mode is offered to help control the power consumption at the expense of GBWP, but with modest results.

MCC API

The MCC provides a complete selection of the operational amplifiers input multiplexing options, automatically enables connected features (i.e. FVR, DAC) and participates to the pin manager allocation of I/Os. A single default OPAx_Initialize() function is generated in the *opa1.c* output file.

PinOut

Operational amplifiers input and output pins are placed at fixed positions to limit size and complexity of the analog multiplexers used. Note that whenever an operational amplifier is enabled, the corresponding output pin is automatically selected and the digital output settings of the pin are ignored.

Homework

- Consider how you could implement a simple auto-calibration (auto-zero?) algorithm to detect and compensate for the OPA input offset.

- Similarly, and possibly with assistance of the DAC and/or the FVR, consider how you could calibrate/compensate for gain errors when using low precision (low cost) external components.

Online Resources

AN1536 - Ultrasonic Range Detection
AN682 - Using Single Supply Operational Amplifiers in Embedded Systems
AN722 - Operational Amplifier Topologies and DC Specifications
AN723 - Operational Amplifier AC Specifications and Applications
AN866 - Designing Operational Amplifier Oscillator Circuits For Sensor Applications

Chapter 8 - Math Support

AT – Angular Timer

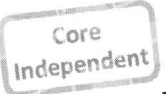

Core Independent

Introduction

In many motor but also AC power applications the control function must be performed synchronously with a periodic signal, be it the AC (sinusoidal) line voltage or the rotation of the motor. In such cases the application of any control algorithm is always making reference to the phase angle, a point in time measured relatively to the beginning of each period. A synchronization input signal is made available (provided by zero crossing detectors, Hall sensors, optical encoders ...) and all timings within the application are constrained by it and by the resulting measured period. But typical microcontroller timers count clock ticks, units of time, only indirectly related to the phase angle. A transformation between the two domains (time - angles) is therefore required. A simple linear proportion determines the relationship between the two domains as expressed in Formula 8.1

a)
$$Phase(rad) = T * (2*PI / T_{period})$$
b)
$$T = Phase(rad) * (T_{period} / 2*PI)$$

Formula 8.1 - Conversion from Time to Phase Angle Domain

Where time measurements are performed in clock ticks or micro-seconds (us) and angles are expressed in radians (replace PI with 180° if degrees are preferred).

The formula has the unpleasant property of requiring the use of a division even when all precautions are taken to reduce the calculations to fixed point math and possibly integer-only math.

When the period can be known (AC applications at 50 or 60Hz frequency), it is common to use look up tables, at the cost of a reduced flexibility and larger use of memory space. Still table size constraints dictate that only a limited number of points is available and rough approximation are required in between.

The Angular Timer removes all such concerns and limitation by providing the ability to generate events (and interrupts if required) directly based on selected phase angle values (angle compare). It can also capture directly the phase angle value of events detected bypassing completely any need to perform the time/angle domain transformation.

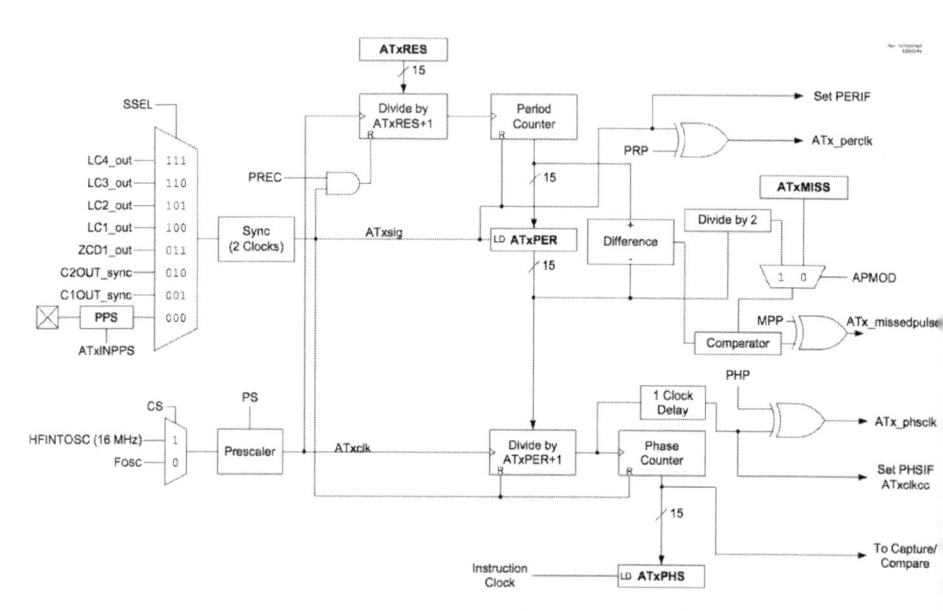

Figure 8.1: Angular Timer Block Diagram

How it works

While complete and relatively detailed the block diagram of the Angular Timer (see Figure 8.1) in itself does not suffice to explain how it works. In the following, I will attempt a more discursive explanation instead.

The Angular Timer is implemented as two nested counters: the *period counter* and the *phase counter*. The period counter is set to reload at the desired *resolution* (if degrees are desired for example, the value 360 can be used). The number of re-loads performed between two consecutive inputs is measured (e.g. 20 cycles of 360 counts). The value obtained (e.g. 20), is a measure of the input period. This is then used as a sort of prescaler value for the phase counter. The phase counter will tick now at the exact desired rate (1 deg per tick in our example producing 360 cycles of 20 counts).

Comparing the main counter value with a number of registers (similarly to what is done in the CCP modules) we can generate directly up to three output pulses or interrupt events that are based on absolute phase angle values. Similarly it is possible to perform up to three captures per cycle, obtaining three absolute phase angle values for the respective events.

In other words, an angular timer module performs the instantaneous time/angle domain transformation at no cost to the CPU, but also provides the equivalent capabilities of three additional capture and compare modules.

Applications

As mentioned in the introduction, it is natural to imagine direct applications of the angular timer module in AC powered applications (home appliances, lighting, AC motor control...) where *phase cutting* techniques are used, possibly in combination with a TRIAC or other power device, to control the load.

The use of the AT will remove most of the application complexity and will make it easier to design solutions that are completely AC-frequency independent.

In motor control applications, the AT makes it easier to define the point of *commutation* independently of the motor rotation speed.

The AT is a relatively new peripheral currently available in the sole PIC16F161x family of devices and therefore its true potential is still vastly unexplored. I am sure I will soon be surprised by the novel and unexpected uses that designers will find for this module.

Limitations

The resolution of the AT module is limited by design to 10-bit, giving approximately 1/3 of a deg resolution or 1024 points along the period. But it is possible that the effective resolution of the application will be limited by the available reference clock (or system clock). The product of the two counters operating inside the AT is eventually dictated by the reference clock frequency. So for example it can be quickly demonstrated that given a 16MHz clock, the maximum resolution (10-bit) is available only for a minimum rotation speed of 260 rpms (16 MHz/1024/60). Below that speed, unless a clock pre-scaler is used (values up to 1:8 are available) the training counter will overflow before a complete rotation of the motor occurs. In such cases the high clock speed required by some of the main application control algorithm elements might conflict with the optimal resolution needs of the timer.

Similarly at high rotation speeds, when the training counter is obtaining a too small value between each input synchronization pulse, the quantization error (1 tick) naturally part of such measurements becomes large compared to the pre-scaling value and will eat away at the actual resolution of the measurements performed. Careful considerations are therefore required in the configuration of the timer for optimal clock speed and resolution of measurement.

MCC API

The MCC dialog window for the Angular Timer is most useful to make sense of the large number of registers required for its working for a grand total of 30 individual 8-bit registers (including all three capture/compares). In fact without it, the AT configuration would be a really daunting task indeed!
Beside the default `ATx_Initialize()` function, the API is then composed of a very thin layer of functions that, despite being mostly self explanatory single-liners, can prove to be extremely helpful setting the resolution and desired compare angles (`ATx_CC1_Compare_SetCount()`...), or getting period and angle (capture) measurements (`ATx_PeriodGet()`, `ATx_PhaseGet()`...).

PinOut

Being offered exclusively on devices featuring the Peripheral Pin Select, there is complete freedom of choice of inputs (synchronization, captures) and output (compares) on any available digital I/O pin.

Homework

- Check out the Missing Pulse feature offered by the AT timer and its potential uses in motor control.
- Consider how you could connect the ZCD module to the Angular Timer and then chain the events automatically to produce a TRIAC firing sequence possibly modulated and timed automatically without any core intervention.

Online Resources

AN958 - Low-Cost Electric Range Control Using a Triac

PID – Math Accelerator

Core Independent

Introduction

It would seem against the very nature of the Core Independent philosophy to provide an 8-bit microcontroller with a math acceleration engine, as if we were to encourage it to compete in software performance against much more powerful adversaries (DSP, 16 and 32-bit microcontrollers...). But that is not the case, this new peripheral must be taken in the right context. The Math Accelerator module is primarily a *PID engine* and PID controllers are today used in many different applications that include motor control, power supplies, and even lighting where 8-bit microcontrollers are used in volumes. In a recent survey, a user contributed PID software library turned out to be the single most downloaded code example from Libstock.com (a popular embedded application sharing site). PID algorithms are also featured in several application notes dating as far back as the beginning of the PIC16 history.

a)
$$\text{Output} = K_p\, E(t) + K_d\, dE(t)/dt + K_i \int E(t)\, dt$$

b)
$$\text{Output}[z] = \text{Output}[z-1] + K_1\, E[z] + K_2\, E[z-1] + K_3\, E[z-2]$$

where:
$$K_1 = K_p + K_d/T + K_i T; \qquad K_2 = -K_p - 2K_d/T; \qquad K_3 = K_d/T$$
and T is the sampling period

Formula 8.2: PID Continuous a) and Discrete b) Representations

In all such applications the PID algorithm is implemented starting from the Laplace transform of its original formulation (see Formula 8.2) which lends itself to be interpreted as a relatively simple sequence of arithmetic operations once the coefficient (Ki, Kp and Kd) are translated in the equivalent K1, K2 and K3. Note also that E[z-1] and E[z-2] are simply the values of the input (Error) at the previous iteration(s) of the loop.

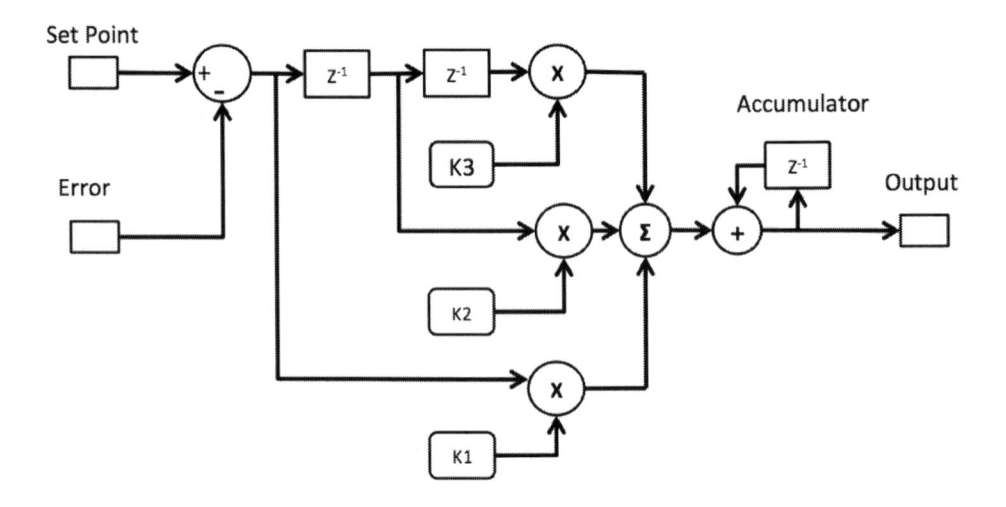

Figure 8.2: Discrete PID Block Diagram

How it Works

Analyzing Figure 8.2, or directly from Formula 8.2, we can see how a PID engine can be assembled by repeated use of an adder and multiplier (A+B)*C using (fixed point 16-bit input and 32-bit output) and an accumulator (with 35 bit capacity).

It was only natural for the PIC16F1 architects to make those operations available independently from the PID function itself. These require only 4 instruction cycles to be executed by the Math Accelerator for a given set of inputs A,B and C.

If the complete PID algorithm needs to be performed, the Match Accelerator is even more efficient, as a total of only 7 instruction cycles are required to perform the entire sequence and, as long as the coefficients don't need to change, only the new Error (input) needs to be fed into the module at each loop.

Applications

Using the Match Accelerator to perform a simple 16-bit addition/subtraction is likely not to be worth the effort as the user needs to take into account the extra instruction cycles spent to load the input registers and fetch the results.

For a 16-bit multiplication things are different. A 16x16 bit signed multiplication with 32-bit result (and accumulation) can save at least a couple of orders of magnitude more cycles (~300 using optimized long arithmetic). A complete PID algorithm approaches the 1,000 instruction cycles when fully implemented in software.

Limitations

The math accelerator is not an integral part of the microcontroller core and therefore the XC8 C compiler won't use the adder/multiplier to accelerate normal arithmetic operations. The user has to *manually* load the peripheral registers of the Math Accelerator and follow the prescribed sequence of operations.

When multiple (nested) control loops are required, the Math Accelerator can still be used to provide a significant boost but the user will have to ensure that not only the coefficients are swapped, but all the intermediate values (z, z-1, z-2) used by the PID engine are saved and correctly restored at each iteration..

MCC API

The MCC takes care of all the tedious work of loading the coefficients and configuring the Math Accelerator in the default `MATHACC_Initialize()` function. During a typical application control loop then, the user will have to provide only updated inputs to the `MATHACC_PIDControllerModeResultGet()` function to retrieve a buffer containing the accumulated output value. Note that since the accumulator is 35 bit large, the result cannot be returned in a `long` type but a special `struct` containing 5 bytes is defined in the *mathacc.h* header file. The user can then choose the best strategy to scale the output value and produce an output (analog/digital) that is appropriate for the application.

Homework

- Compare the speed of the Match Accelerator with the conversion speed of the ADC. Which one of the two is most likely to become the system bottleneck in the example above assuming a 32MHz clock?
- Compare the Math Accelerator performance with that of a basic dsPIC33F or PIC24H device when executing a software PID.
- Estimate the effective number of cycles required to perform a multiply and accumulate operation (including register loading/result unloading).

- Could you use the Math Accelerator in a digital filer (FIR, IIR...) application?

Example

The following example represents a simplified application main loop where a PID algorithm receives input from the ADC and actuates the output with a PWM:

```
/* Project: MathAcc Example
 * Device: PIC16F1619
 */
#include "mcc_generated_files/mcc.h"
#include "mcc_generated_files/adc1.h"        // defines Sensor channel

void main(void)
{
    int SetPoint = 0x1234;
    MATHACCResult Output;
    SYSTEM_Initialize();    // initialize device, MATHACC, ADC and PWM3

    while (1)
    {
        int CurrentValue = ADC1_GetConversion( Sensor);

        Output = MATHACC_PIDControllerModeResultGet( SetPoint,
                                                    CurrentValue);

        PWM3_LoadDutyValue( Output.byteHH + (Output.byteU<<8));
    }
}
```

Online Resources

AN937 - Implementing a PID Controller Using a PIC18 MCU
AN964 - Software PID Control of an Inverted Pendulum Using the PIC16
AN1138 – A digital Constant Current Power LED Driver (using PID)

More Math Related Application Notes:

TB040 - Fast Integer Square Root
AN1061 - Efficient Fixed-Point Trigonometry Using CORDIC Functions
AN821 - Advanced Encryption Standard Using the PIC16
AN942 - Piecewise Linear Interpolation on PIC12/14/16 Series Microcontrollers

Chapter 9 - eXtreme Low Power

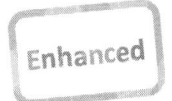

Enhanced

XLP – eXtreme Low Power

As 8-bit PIC microcontrollers have become ever more capable, their power consumption has been squeezed to *extreme* levels. While not all applications need to operate from a battery for a decade straight as some marketing types would like you to believe, there are many advantages that can come from a more conservative use of energy. Smaller power supply circuits, less noise, less heat and hence smaller packages, and in general a small application form factor are all desirable features.

About a decade ago all PIC microcontroller designers committed to a common set of power consumption target figures called the *eXtreme Low Power* standard. The original set included an ambitious maximum current consumption of the device when in Sleep (lowest power mode) of 100 nA (that is nano Amperes!) But also the Watchdog and the Secondary Oscillator (SOSC) had a target of just 800nA each.

Today, all PIC16F1 devices routinely pass those criteria and by a wide margin. In fact it is common to find on the device datasheet values that are almost an order of magnitude lower: 20nA in Sleep and less than 300nA for WDT and SOSC.Even the dynamic power consumption of the microcontroller, that is the current consumption when actually executing the application, has been similarly reduced by orders of magnitude with current values commonly in the 30 uA/MHz (micro Ampere per MHz of system clock).

These are values that would be very hard (if not impossible) to achieve with larger (16 and 32-bit) architectures because they require the use of much smaller, and therefore leaky, CMOS processes. In fact it is common for those architectures to adopt radical expedients (deep sleep modes) just to get near much less ambitious figures (few micro Amperes) at the cost of RAM contents loss and lengthy wake up times. None of that is necessary with the PIC16F1 microcontrollers as even when in the lowest power mode they do preserve the full contents of the RAM memory and yet can wake up in microseconds.

Low Power Modes

Enhanced

Sleep

Sleep has been the signature low power mode of all PIC microcontrollers from the beginning of time. Initially it was used to indicate a mode where the main system clock is stopped and with it pretty much any activity of the microcontroller save for the Watchdog timer thanks to its independent oscillator.

In the PIC16F1 generation of devices though, the list of independent oscillators available on chip has grown considerably and with it the number of possibilities. Sleep still means stopping the system clock, and therefore the core instruction execution, but many peripherals are capable of continuing their operation if configured to use the alternate oscillators. Examples of such peripherals include:

- Timer1, a 16-bit timer, when using the Secondary Oscillator (SOSC) or when used as a counter in asynchronous mode.

- Analog to Digital Converter, when using the dedicated internal oscillator (FRC)

- All core independent peripherals when using directly one of the oscillators (for example HFIntOsc instead of Fosc/4).

Low Power Sleep

The low power sleep feature must not be confused with the *deep* sleep mode of other architectures. It applies only to PIC16F1 devices that operate at 5V and therefore using the internal LDO. Such LDO can be switched to a lower power mode (with resulting lower quiescent current) when selecting the low power feature. Note that RAM contents integrity is fully preserved, but wake up time is affected as the return to full power requires a short stabilization delay.

Wakeup

There are several events that can force the microcontroller to exit the low power state and resume operation, including:

- External Reset (MCLR)

- BOR reset

- Watchdog timer
- Any external interrupt
- Any interrupt produced by a peripheral that is operating asynchronously with the system clock

Regardless of the trigger event, waking up from sleep, the MCU will continue executing the instruction immediately following in program memory.
This is true even for peripherals that would otherwise generate an interrupt if the interrupt enable bit is cleared (not enabled).

Idle [New]

Idle mode was introduced first in PIC18 models and more recently in PIC16F1 models (such as the PIC16F183xx and PIC16F188xx families) to allow the microcontroller to stop executing while keeping all other clock systems up and running, including those peripherals that are configured to operate off the system clock. The power saving achievable is noticeable but not remotely comparable to the savings achieved by Sleep mode.

Doze [New]

Doze mode was similarly introduced in the most recent PIC16F1 families to further increase the granularity of power consumption control. When in Doze mode the core is allowed to continue to run although its clock source is divided further by a factor that can be selected between 1:2 to 1:256.
The result is once more a power consumption reduction proportional to the scaling factor chosen, with the benefit of allowing all peripherals to continue operating as well as some (reduced) core activity.

Doze Interrupt Boost [New]

When in Doze mode, it is possible to configure the device so that an incoming interrupt will make the core return instantaneously to full speed. It is possible then to ensure that the core will return to doze mode upon interrupt exit. Effectively the combination of the two options can be used to produce a sort of interrupt boost. In other words, it is possible to conceive low power application where the main application loop is normally executed at a desired fraction of the clock (i.e. 1:8 = 4MHz) and only interrupt routines get to use the full speed and maximum performance the microcontroller is capable of (1:1 = 32MHz).

PMD – Peripheral Module Disable *New*

The power consumption of a peripheral module that is *not used* by an application is typically very limited, but can be completely eliminated by *disabling* entirely the peripheral via the *PMD* control register available on the very latest generation of PIC16F1 devices (PIC16F183xx and PIC16F188xx for example).

A peripheral that is disabled does not receive anymore any of its clock and input stimuli. Its registers are not available anymore to the microcontroller core and all I/Os eventually assigned to it are released and returned to the next peripheral/option according to a pin priority.

When re-enabled, the peripheral state will be returned to the POR reset condition as documented in the device datasheet.

Achieving XLP

Achieving eXtreme Low Power consumption in an application requires much more than simply entering Sleep mode. Here is a short list of items that the designer should check, and consider turning off before entering Sleep mode, as they can impact significantly the resulting power consumption of the device:

- I/O pins left floating
- External circuits sinking current from I/O pins
- Weak pull-up option if enabled
- Modules using LFintOSC (i.e. Fail Safe Clock Monitor)
- Core Independent peripherals that are left connected directly to oscillators (i.e. CWG or NCO and HFIntOSC)
- Analog modules that make use of the internal voltage reference (FVR) when configured to use it (i.e. DACs, BOR, OPA, Comparators...)

MCC API

MCC makes it easy to generate multiple initialization functions for each module. For example, beside the default initialization that will be used during the `SYSTEM_Initialize()` sequence, we can create a low power configuration to be used before entering Sleep mode.

Example

Create a new initialization function for the FVR module, call it "Sleep" and set it so to turn off the entire module:

```
// < MCC generated code >
void FVR_Sleep(void)
{
    // CDAFVR off; TSRNG Lo_range; TSEN disabled;
    //   FVREN disabled; FVRRDY disabled; ADFVR off;
    FVRCON = 0x00;
}
// < MCC generated code >
```

Have one of these for each module that could unnecessarily keep your application power consumption up and call them before going to sleep.

```
// turn all things off (but the chosen wake up)
FVR_Sleep();
ADC_Sleep();
// ...

// go to lowest power mode
CLRWDT();
SLEEP();
NOP();
```

Homework

- True low power design requires a well thought out plan. To get a clear picture of all the inter-dependencies between modules and to find the best possible configuration for your application, always check the "Operation in Sleep" section of each peripheral in the device datasheet.
- Compare the wakeup capabilities of the Interrupt on Change feature vs. the CLC interrupt when used to detect external (digital) events.
- Always refer to the datasheet for *maximum* values of power consumption in the application temperature and voltage range.
- Read the excellent Low Power Report from Jack Ganssle (see link below) which dispels the many myths about ultra low power applications and provides excellent guidelines for reliable (and realistic) performance/long battery life.

Online Resources

http://www.ganssle.com/reports/ultra-low-power-design.html

AN1416 - Low-Power Design Guide
AN1337 - Optimizing Battery Life in DC Boost Converters Using MCP1640
AN1088 - Selecting the Right Battery System for Cost-Sensitive Portable Applications While Maintaining Excellent Quality
AN1288 - Design Practices for Low-Power External Oscillators

Appendix A – Peripheral Integration Guide

Product Family	Pin Count	Program Flash Memory (KB)	ADC (# of bits)	Comp	HSComp	DAC (# of bits)	HC I/O (mA)	OPA	PRG	SlopeComp	ZCD	CCP/ECCP	10-bit PWM	16-bit PWM	PSMC (16-bit PWM)	COG	CWG	NCO	DSM	AngTMR	HLT (8-bit)	PSMC (16-bit)	16-bit PWM (16-bit)	NCO (20-bit)	SMT (24-bit)	RTCC	TEMP	CLC	MULT	MathACC	CRC/SCAN	HLT	WWDT	EUSART/AUSART	I2C™/SPI	USB with ACT	LIN Capable	mTouch® Sensing	HCVD	LCD	HEF	PPS	IDLE and PMD	DOZE	XLP
PIC10(L)F3XX	6	384–896 B	8	✓									✓				✓	✓						✓			✓	✓										✓							✓
PIC16F527/570	20–28	1.5–3	8	✓																								✓										✓						
PIC1X(L)F150X	8–20	1.75–14	10																									✓								✓	✓		✓						
PIC16(L)F151X/2X	28–64	3.5–28	10			5						✓	✓															✓						✓	2	✓		✓			✓				
PIC12LF1552	8	3.5	10																																✓	2			✓	✓	✓	✓				
PIC16LF1554/9	14–20	7–14	Dual 10	✓										✓	✓									✓					✓						✓	✓			✓	✓	✓	✓				
PIC16LF145X	14–20	14	10	✓		5							✓	✓															✓						✓	✓	✓		✓	✓	✓	✓				
PIC16(L)F157X	8–20	1.75–14	10	✓												✓	✓	✓										✓						✓	✓		✓	✓	✓						
PIC1X(H)VF762/63	8–14	1.75–3.5	10	✓	✓	5/9	50	✓		✓		✓	✓			✓				✓	✓						✓	✓			✓	✓	✓	✓	✓		✓	✓				✓			
PIC16(L)F182X/4X	8–20	3.5–14	10	✓		5							✓	✓															✓					✓	✓	✓		✓	✓		✓	✓	✓			
PIC1X(L)F1612/3	8–14	3.5	10	✓													✓											✓	✓						✓	✓		✓	✓	✓				✓			
PIC16(L)F161X	14–20	7–14	10	✓		8	100							✓		✓	✓	✓	✓		✓			✓	✓		✓	✓			✓	✓	✓	✓	✓		✓	✓	✓				✓			
PIC16(L)F170X/1X	14–40	3.5–28	10	✓		8		✓				✓	✓	✓			✓					✓						✓	✓						✓	2		✓	✓							
PIC16(L)F176X/7X	14–40	7–28	10	✓		5/8		✓				✓	✓	✓			✓					✓						✓	✓						✓	✓		✓	✓							
PIC16(L)F178X	28–40	3.5–28	12	✓		5/10	100	✓	✓	✓	✓	✓	✓															✓						✓	2		✓	✓							
PIC16(L)F183XX	8–20	3.5–14	10	✓		5/8		✓				✓	✓				✓					✓						✓	✓					✓	✓	✓		✓	✓		✓		✓	✓		
PIC16(L)F188XX	28–40	7–56	10*	✓		5						✓	✓	✓												✓	✓	✓			✓			✓	✓	2		✓	✓	✓	✓		✓	✓	✓	✓
PIC16(L)F193X/4X	28–64	7–28	10	✓									✓	✓															✓						✓	✓		✓	✓	✓				✓	✓	✓
PIC18LFXXK20	28–40	8–64	10	✓									✓	✓																					✓	✓								✓		
PIC18(L)FXXK22	20–80	8–128	10	✓		5							✓	✓				✓	✓										✓						✓	2		✓	✓					✓	✓	✓
PIC18(L)FXXK40	28–64	16–128	10*	✓		5						✓	✓	✓												✓		✓	✓						✓	2		✓	✓	✓				✓	✓	✓
PIC18(L)FXXK42	28–64	16–128	10*	✓		5						✓	✓	✓				✓	✓							✓		✓	✓					✓	✓	2		✓	✓	✓				✓	✓	✓
PIC18(L)FXXK94	64–100	32–128	12	✓									✓	✓												✓		✓	✓	✓		✓	✓	✓	✓	4	✓	✓	✓	✓		✓		✓	✓	✓
PIC18(L)FXXK50	20–40	8–32	10	✓									✓																✓						✓	✓	✓		✓						✓	✓
PIC18FXXK90	60–80	32–128	12	✓									✓													✓			✓						✓	2			✓		✓			✓	✓	✓

Appendix B – PIC16F + 2 and 3-digit Decoding Ring

The part numbering scheme of PIC16 microcontrollers is a mystery to most but the closest circle of initiated. While some believe that most of the part numbers had been generated using a powerful Random Number algorithm, or perhaps that Keeloq encryption algorithms had been employed, I created the illustration below to try and dispel such myths.

Notice that this decoding ring is applicable (not without exceptions) to PIC microcontrollers of the PIC16 series that were introduced between 1989 and 2014.

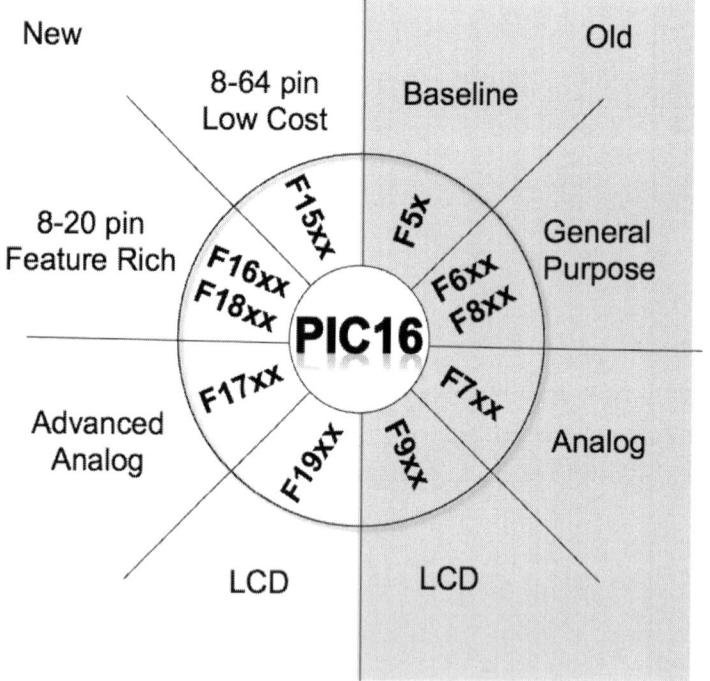

Parts marked PIC16F5x feature the original PIC Baseline core.
Parts marked PIC16F plus three digits feature the Mid Range core.
Parts marked PIC16F1 plus three digits feature the Enhanced Mid Range core and a growing number of Core Independent Peripherals

Appendix C – PIC16F1 + 4 digit Decoding Ring

As of late 2014 with the number of individual devices in the PIC16 family passing the 1,000 mark, a new part numbering scheme has been adopted. This is distinguishable by the presence of up to 5 digits total.

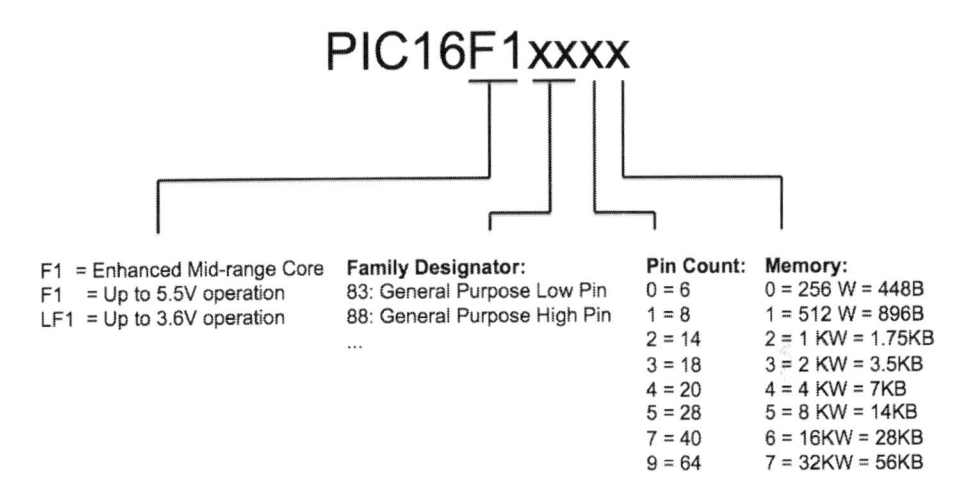

$$PIC16F1xxxx$$

F1 = Enhanced Mid-range Core	**Family Designator:**	**Pin Count:**	**Memory:**
F1 = Up to 5.5V operation	83: General Purpose Low Pin	0 = 6	0 = 256 W = 448B
LF1 = Up to 3.6V operation	88: General Purpose High Pin	1 = 8	1 = 512 W = 896B
	...	2 = 14	2 = 1 KW = 1.75KB
		3 = 18	3 = 2 KW = 3.5KB
		4 = 20	4 = 4 KW = 7KB
		5 = 28	5 = 8 KW = 14KB
		7 = 40	6 = 16KW = 28KB
		9 = 64	7 = 32KW = 56KB

Notice that, the since counts of 6 and 8 pins are now normally included in the fourth digit of the part number, the use of the PIC10 and PIC12 prefix has been discontinued.

The new scheme is supposed to eliminate much of the guesswork to determine pin count and memory sizes of devices. A two digit numerical *family designator* code remains and is *likely* to continue carrying some resemblance with past generations (9 LCD, 7 Analog, 5 Entry Level, 6 and 8 general purpose...).

Alphabetical Index

8 and 16-bit Timers...35
Active Clock Tuning...102
ADC – Analog to Digital Converter..116
ANSEL – Analog Select..62
AT – Angular Timer..127
BOR – Brown Out Reset..89
CCP – Capture Compare and PWM...39
CLC – Configurable Logic Cells...64
COG – Complementary Output Generator..44
Comparators..107
Core Independent Peripherals..4
CRC – Cyclic Redundancy Check with Memory Scanner......................................83
CWG – Complementary Waveform Generator...44
DAC – Digital to Analog Converter...110
Data EEPROM...77
Doze..137
DSM – Data Signal Modulator...71
EUSART – Asynchronous Serial Port...98
Flash Memory...78
FSCM – Fail Safe Clock Monitor..90
FVR – Fixed Voltage Reference..113
HCVD – Capacitive Sensing..119
HEF – High Endurance Flash...79
HIDRV – 100mA..63
High Speed Comparators...109
HLT – Hardware Limit Timer...53
I2C – Inter Integrated Circuit Bus..91
Idle...137
INLV – Input Level..63
Installing MPLAB X...11
IOC – Interrupt On Change..63
LAT – Output Latches...62
LPBOR – Low Power BOR...89
MCLR – Master Clear...90
MLA – Microchip Library for Applications – Touch Library.................................120
MPLAB Code Configurator...14
MPLAB X...11
    Header Files folder...18
    Logical Folders...18
    New File Wizard...19
    New Project wizard..16
    Source Files Folder...18
MPLAB XC8...12
NCO – Numerically Controlled Oscillator...48
ODCON – Open Drain Control..62
OPA – Operational Amplifiers...124
Oscillators...31

PICkit™ 3 8
PMD – Peripheral Module Disable 138
POR – Power On Reset 89
PORT – Direct Pin Access 61
PPS – Peripheral Pin Select 68
Reset Circuits 88
Sleep 136
SLRCON – Slew Rate Control 63
SMT – Signal Measurement Timer 57
SPI – Synchronous Port 94
Temperature Indicator 122
TRIS – Tri-state control 61
USB – Universal Serial Bus 102
Wakeup 136
WDT – WatchDog Timers 86
WPU – Weak Pull Ups 62
XLP – eXtreme Low Power 135
ZCD – Zero Cross Detect 73
   52p., 119